How We Make Ducks Pay
Profitable Breeding of the Modern Pekin All White Mammoth Ducklings

by American Pekin Duck Company

with an introduction by Jackson Chambers

This work contains material that was originally published in 1906.

This publication is within the Public Domain.

This edition is reprinted for educational purposes
and in accordance with all applicable Federal Laws.

Introduction Copyright 2017 by Jackson Chambers

IMPORTANT NOTE & DISCLAIMER

IMPORTANT NOTE :
As with all reprinted books of this age that are intended to perfectly reproduce the original edition, considerable pains and effort had to be undertaken to correct fading and sometimes outright damage to existing proofs of this title. At times, this task can be quite monumental, requiring an almost total rebuilding of some pages from digital proofs of multiple copies. Despite this, imperfections still sometimes exist in the final proof and may detract slightly from the visual appearance of the text.

DISCLAIMER :
Due to the age of this book, some methods or practices may have been deemed unsafe or unacceptable in the interim years. In utilizing the information herein, you do so at your own risk. We republish antiquarian books with no judgment or revisionism, solely for their historical and cultural importance, and for educational purposes.

Self Reliance Books

Get more historic titles on animal and stock breeding, gardening and old fashioned skills by visiting us at:

http://selfreliancebooks.blogspot.com/

INTRODUCTION

I am very pleased to present to you another important old poultry publication – *How We Make Ducks Pay*. It was written by the *American Pekin Duck Company*, and first published in 1906, making it over a century old.

Breeding your own animals and growing your own food is on the rise right now. More and more people are turning away from store-bought foods whose ingredients can be dubious at best, and toxic at worst, and raising their own poultry for meat and eggs.

If you've been pondering whether or not to take the plunge, maybe this book can help you make your decision. It is a great starting point for those even considering a poultry operation, and will give you a glimpse into what is required to run a successful duck farm.

The book features chapters like *Ducks for Business, Shelters & Ventilation, Care of Breeding Stock, Sex, Pairing & Breeding, Fattening, Killing, Picking & Shipping*, and more.

This book may be a centenarian, but the straight-up, no-nonsense advice and tips on raising ducks are timeless. Good, solid knowledge never gets old!

Jackson Chambers

State of Jefferson, November 2017

AMERICAN PEKIN DUCKLING

One of our birds, a female, ten weeks old, saved for a breeder and with some of her fat trained off by running around the range. The parallelogram shape of body which is aimed at in breeding is here in evidence.

This is one of several poses which a duck assumes. She is in repose. At feeding time, and when they are waddling, the necks of both ducks and drakes are elongated and the body is tipped upwards, the whole poise being different from what this picture shows. Many birds are so fat at killing age that their bellies rest on the ground when they are standing in repose like this bird. When such birds walk, they have to make an effort to pull their bellies off the ground, and roll from side to side in their walk, like a sailor ashore.

CONTENTS

		Page
	Introduction	9
I.	Ducks for Business	11
II.	Shelter, Ventilation	23
III.	Care of Breeding Stock	29
IV.	Sex, Pairing, Breeding	39
V.	The Egg	47
VI.	Youngest Ducklings	55
VII.	Fattening	69
VIII.	Killing, Picking, Shipping	75
IX.	Markets	91
X.	Question Box	109

UNDER THIS SHELTER ROOF TO GET OUT OF SUN AND RAIN

LIST OF ILLUSTRATIONS

	Page
American Pekin duckling (*Frontispiece*)	
Under this shelter roof to get out of sun and rain	7
Here come the little ones out of the nursery for an airing	14
Ducklings five weeks old in yards of second nursery house	18
Feeding time for ducklings in the yards of the fattening sheds	20
First nursery yards	22
Exterior of brooder house	24
Interior of brooder house	26
On grass	30
Food-mixing machine	34
Drake, showing curl feather	40
How to handle live birds	44
Where they lay their eggs	48
Incubator cellar, showing machines which hatch the eggs into ducklings	52
Egg-testing lamp	53
Fertile egg, seventh day	53
Ducklings two weeks old	56
Water fountains and food boards	58
Thousands of youngsters	60
Ducklings in the sun at mid-day	64
Double brooder house	66
Winter scene, brooder house	68
Ready for feeding	70
Meal time	72
Fattening yards	74
Sticking knife	75
Knife used in picking	77
Duck picker at work	78
Ready for shipping	82
Warm house, and yards	88
Pen of ducklings on dirt run a week previous to killing	90
Parallelogram body, depth of keel and plumpness	112
Fattening shed and pens	118
Shelter roof for ducks on the range	121

INTRODUCTION

The breeding of our modern large Pekin duckling is an industry whose possibilities a few of us have known for a long time, and demonstrated to the best of our ability. We believe many will be interested in our methods. We know thousands will see a light in these pages which will astonish them, as our own work has astonished us. The strength of ducks as a business deserves telling,—the whole truth and nothing but the truth.

It is difficult for us to work day by day and at the same time teach the business to beginners and others in a convincing and satisfactory manner, without a book like this. The answering of hundreds of letters weekly from those asking for advice and breeding stock is a tedious, incomplete and poor method of telling the hows and whys.

We offer this instruction book to the American people and all nations in the hope that we are doing this branch of the great poultry industry a real and lasting service. We know the country is just getting awakened to the idea.

The aim in writing it has been to give the details, and all the details, with the fullness, clearness and simplicity which a man or woman investigating a new subject demands. We give them in the order in which the subject naturally presents itself.

We have tried to keep out as much as possible from this work our own enthusiasm over the duck business, intending to present the facts calmly and soberly so that the strength of the proposition would follow as a matter of course, without exaggeration. At the same time, we would not undervalue a proper amount of enthusiasm. He who would make the most out of ducks must feel a lively interest in the work,—that it is paying work,—work which will reward him surely and amply, as it has rewarded us. We would advise the beginner, if he decides to take hold, not to go ahead in a listless, doubting, skeptical way, but confidently, knowing that he is able, and can do what others have done. There are no difficulties in the duck business which need appal the most timid, nor can great financial loss overtake one who observes common prudence.

In this beginning, we must acknowledge the obligations which the modern duck industry owes to Mr. James Rankin for his life-time of labor in it. Looking back serenely upon a past fruitful with good works,—and enjoying a fortune which he, as well as others, has made in the breeding of ducks,—but now beyond the age of threescore years and ten, this pioneer is giving the standard up to others to be carried. He has always been our friend; we wish him continued life, health and prosperity in full measure.

CHAPTER ONE

DUCKS FOR BUSINESS

The Pekin is queen.—Other varieties and their characteristics, briefly summarized.—Duck raising worth all one's time and attention, because it is profitable in proportion to the number raised.—Our ducks have web feet, but not the desire to swim.—Pond not needed.—Ducklings ready for market in only ten weeks.—Cost of raising them six to ten cents a pound, including labor.—Profit of fifty cents a duckling to be figured.—Ducks have no lice, no diseases, and hawks will not carry off the young.—Food they eat is the cheapest fed to any live-stock.

The white Pekin is the queen of all ducks. This breed was brought to America from China, where they are raised now, and highly esteemed. The first specimens of Pekins were imported fifty years ago. In this comparatively short period, a great deal has been done to improve the size and breeding qualities of this variety.

Other varieties of tame ducks are the white **Aylesbury, Indian Runner,** white **Muscovy,** colored **Muscovy,** colored **Rouen,** black **East India,** black **Cayuga,** blue **Swedish, Crested White,** gray **Call,** white **Call.**

Everything written in this book applies to the white Pekin strain which we have developed. These teachings do not apply to the other above-named varieties. Commercially, we believe in the Pekin only. The other varieties may be bred for show-room or for amusement, but as to their money-making qualities we are in doubt. **We have no doubts as to our Pekins.** They are easily raised in great numbers and are in active demand in the markets, and sold at a large profit.

America leads in the development of the Pekin. The duck breeders in England are few and far between. They have experimented most with the Aylesburys. The Pekins mature more quickly than the Aylesburys and their feathers come out more easily when picking. The Indian Runner duck is smaller than the Pekin and has dark pin-feathers. It is not so salable as the Pekin; side by side in the markets the Pekin will beat it in looks, size and price. The Indian Runner is a good layer but this does not offset the other points mentioned. Muscovy ducks are fighters and hard to handle. Gray and white Call ducks are bantams bred for the show-room but never for profit. The black East India is another bantam variety, seldom weighing more than two pounds. The other above-named varieties are rare.

In disposition, as well as size, quick maturity and fecundity, the Pekin is exceptional. Pekins do not quarrel. They are easily driven and handled.

Duck raising should be entered into not for amusement or for a pastime, but **to make money.** It is worth all of one's time and attention and will richly reward the earnest, methodical worker. The main thing is to know how to do the work and then do it. We have tried to make this book so simple and plain that there can be no question as to procedure. There are breeders of ducks now in many parts of the country who have been working with poor stock, and largely by guesswork, and what they could learn by experience. Comment on duck raising as given in the periodical press from time to time is for the most part absurd, written by inexperienced experimenters groping in the dark. Many of these experimenters are breeding the common, or puddle ducks, or the light-weight ducks of colored plumage, all of inferior size and fed on lake or sea-shore fish until the flesh tastes more or less fishy.

Our strain is as different from the common ducks as day is from night. Our birds are what we have made famous as the cross-bred, white-feathered Boston ducklings, fattened on grain and beef scraps, weighing five to six pounds when marketed at ten weeks of age. We have

DUCKS FOR BUSINESS

shipped this breeding stock to beginners as well as experienced breeders, and we find that a profitable success is made with them.

We do not ask anybody to try what we have not done ourselves. Ducks have been a gold mine for us. We have done it as the ordinary breeder would do it, with common, plain work, shipping to an every-day market. For ten years we have been of the opinion that the advantages of this business for the good of all should be made known, and we are ready to do our best in this line now. We ask a hearing by calm, sober, common-sense people. We stand behind all statements we make, and **guarantee that they are the simple truth.**

Perhaps the most surprising point, to the average reader, is that **our ducks are raised without water. Don't think you must have a pond or brook on your place.** A farm which is good for nothing from the farmer's standpoint is just the place for ducks. The manure will make the most sterile fields productive enough for the green stuff and vegetables that may be grown.

Our ducks have web feet but we have bred out of them the desire to swim and bathe. The advantages of no swimming water were made manifest to us quickly. We know the methods and flocks of a few duck breeders who use water and we never could find that the ducks were better in any way.

Perhaps you have a brook or pond. A brook is handy, perhaps, in that it will reduce the work of watering. But the ducks will not get any larger or fatter because of it. They will drink from a brook or pond, thereby lessening the work of their caretaker.

Anybody raising ducks with a pond on his place will find that some ducks **will lay eggs in the water.** Unless the water is shallow and the eggs easily reached, this will be a source of annoyance and loss. As ducks lay at night, or early in the morning, this trouble can be overcome by shutting the birds up at night and not letting them into the water until about 9 a. m.

HERE COME THE LITTLEST ONES OUT OF THE NURSERY FOR AN AIRING

DUCKS FOR BUSINESS

If you have a spring or brook with a fall so that water can be diverted and made to run through the duck house or houses, that may be worth trying.

Most beginners without instruction think that ducks must have swimming water to thrive and, lacking a pond or brook, will dig a rainhole without inlet or outlet. This quickly gets muddy and slimy and becomes an abomination, a menace to both ducks and owner.

We have heard of duck raisers on the coast of Delaware, who have had trouble in the following way: The tides would force the stagnant marsh water back into the duck ranges and when the ducks got into this brackish water it was bad for them. Some actually would be poisoned and die. Look out for this stagnant, foul-smelling marsh water if you breed ducks on the sea-coast.

The rice-fields of the South are ideal for ducks. They will pick up plenty of free nourishment there. Ducks are good also to go over harvest fields to pick up the left-behind grain. This is a suggestion for large farmers who breed ducks.

Our ducklings are not raised as a small breeder raises chickens. An **incubator** is the device which multiplies the money-making possibilities. In the first place, these modern ducks will not reproduce their young by nest-building and setting. They have been bred to lay eggs, and not to set on them. If you wish to start small and without an incubator, you have got to take a common, old-fashioned hen and set her on the duck eggs to hatch them out. On account of the incubators we save great expense by carrying only comparatively few breeding birds from one year to the next. You see, **from each duck we get eggs enough to raise three-score ducklings that year. Every duck worth five dollars to ten dollars reproduces young worth three hundred dollars to six hundred dollars** (if saved for breeders). If killed and sold at market they are worth three-score times the market price of the one duck which started the big family. **These figures are not dreamed. We are doing it right along.**

HOW WE MAKE DUCKS PAY

A duck reproducing at that rate must be good to start, and must have intelligent care and feed.

Ducklings are on your hands only ten weeks. Then they vanish to market and you get your money for them, and their room also. This goes on day after day, raising, killing, shipping. There is no let-up until you get ready, and the markets take them all the time, as they do chickens. Prices for ducklings are highest in the East in April and May. It is not necessary to get these high prices all the year to make the business a success. **We have taken the markets as they have come every month of the year, knowing all the while we were making a good profit,** even when prices were at times one-half to one-third lower than at others.

We know by experience the cost of raising ducklings (food and labor including expert pickers making from twenty dollars to thirty dollars a week), to be from SIX TO TEN CENTS A POUND, depending on the fluctuating prices of grain. Others say the same, including the government reports from experimenters; so the reader does not have to take our word alone for this fundamental expense.

It costs, therefore, from thirty to sixty cents to get the duckling up to the market and into the market. The selling price is fifty cents more than this. For many years after learning the business by hard knocks, we have figured confidently on making half-a-dollar profit on every duckling,—that is what actually happens.

That is what it will do for you, starting with our stock and following our instructions. There is no legerdemain about it. Nor is it ridiculously easy — do not believe that.

Anybody can start with half a dozen ducks and drakes and not have a busy working day, exactly the way a few hens may be kept by a man working eight hours daily at something else.

You can demonstrate that they are money makers with a few of them. To get an income of twenty dollars a week you must send to market forty ducklings a week; to

DUCKS FOR BUSINESS

make your farm produce a profit of one hundred dollars a week, you must send to market two hundred ducklings a week.

The cost of from six to ten cents a pound to bring a duckling to killing age, according to the location of the plant and according to the prices of grain, **includes labor as well as food.** Figuring food alone, **five cents a pound** would cover the cost.

Of course a duckling does not eat so much when it is small and newly hatched as it does when it has reached the killing age. If you keep a duck from killing age on for breeding it will cost you about twenty-five cents a month in feed and labor to carry the duck. **This is why good breeding stock sells for much higher prices than the killed ducklings.**

The labor charge is cut down in proportion to the increase in the number of ducks kept. The care of thirty thousand ducks may be divided among six men.

Bearing in mind what we have said about cost it may be estimated accurately that a duckling of market age, weighing five pounds to six pounds, will cost to produce from thirty to sixty cents.

The wholesale selling price is at least twelve cents a pound, depending on the market and the season. Twelve cents is the lowest we have ever known it here, and thirty cents the highest. This means that each duck will be sold for sixty cents to one dollar and fifty cents. A duck which has cost the high price to produce will sell for the high market price; for this is the way the market runs. This means that a profit of at least fifty cents a duck is going to be made. Everything which we ever read on ducks confirms our own experience in the matter of profits, and it will be found invariably that the breeder shipping ducks to market, even on a small or large scale, makes this profit. The amount of profit depends on the number of ducks handled.

What we say about the cost, selling price and profits is positively and absolutely true. We hold no brief to boom

DUCKLINGS FIVE WEEKS OLD IN YARDS OF SECOND NURSERY HOUSE

DUCKS FOR BUSINESS

the duck industry. We insist respectfully and most earnestly that we have made, and are now making, the profits which we say are in ducks, selling to market only; and what we have done and are doing, others have done and are doing. Our statements will find ready corroboration from any duck breeder who is really breeding ducks, that is to say, who is actually turning them out and shipping them to market, and not playing with a small flock for fun, or secondary to other work.

In speaking of these profits, we do not estimate the sale of breeding stock. If you keep what you raise until they are of breeding age, and then sell them to your neighbors, or to anybody, by advertising or exhibition, you will make more. Nor do we take account of the sale of duck eggs. Duck eggs are salable on account of their large size and good cooking qualities, and many are in the markets, but the big duck raiser has a better use for most of his eggs than the table; he has his incubator in mind. He wants them for seed.

Our ducks have no lice or other vermin. They are not bothered by hawks. They have no diseases. Good, strong points, those three. Lice are a terrible nuisance in general poultry raising and have dissipated many a dream of profits. When we say that our ducks have no vermin on their bodies of any kind, we mean just that, **and we say it emphatically.** Why it is exactly, we do not know. One would imagine that a louse would live on any animal. Still, there are others (rabbits, for instance) which have no vermin.

Hawks are an ever-present pest in many parts of our country. They will not touch the youngest duckling.

By no disease, we do not wish to give the impression that ducks resist all ill-treatment. Fed improperly, they will have diarrhœa. Kept in the sun constantly when little and given no shade, they will be sun-struck. Given a chance when very young to eat certain bugs, they will be killed. Allowed to become crowded and panic-stricken, they will get lame and otherwise injured.

FEEDING TIME FOR DUCKLINGS IN THE YARDS OF THE FATTENING SHEDS

DUCKS FOR BUSINESS

Starved, they will die like any animal. But these matters are **absolutely under the control of the breeder,** with very simple and sure arrangements. **There will be no losses from what is commonly known as disease.** No medicines or pills or drugs of any kind are of any use in the duck business.

The beautiful **white feathers** picked from the ducklings before marketing are worth good money, **forty-five to fifty cents a pound.** Every twelve ducks will give a pound. Generally this revenue will pay the picking.

The following food is given (See further chapters in this book for details.): **Rolled oats, bread-crumbs, bran, corn-meal, flour (low-grade), beef scraps, green stuff, vegetables, grit, ground oyster shells.**

The **rolled oats** cost in New England $3.25 to $5 a barrel (one hundred and eighty pounds). This is the most expensive item in the ration. They are fed only to the youngest ducklings and to them only a brief period.

The **bread-crumbs** are made from stale bread given away (or sold for little) by bakers. Bought in quantities it costs only twenty-five dollars a ton.

The **bran** (also called shorts) is the outside shell or wrapper of wheat. It costs only twenty dollars a ton but in the West near the flour mills it is much cheaper.

The **corn-meal** is common yellow Indian meal which has been ground (not cracked). It costs from $1 to $1.25 per one hundred pounds in the East; in the West it is cheaper.

The **low-grade flour** costs twenty-eight dollars a ton in the East; cheaper in the West.

The **beef scraps** cost $2.50 per one hundred pounds. They form a small per cent of the ration, at a certain time.

By **green stuff** is meant anything growing, like common grass, oats, clover, rye, millet, etc.

The **vegetables** are cheap on the average farm, and are a fine duck food. Turnips and carrots are easily raised and turned into duck meat.

The **grit** may be ordinary sand or gravel for a certain

HOW WE MAKE DUCKS PAY

period in the life of the ducks, after which the **cheapest granite grit** is bought. Fancy grits costing as much or more than grain are not used.

The **ground oyster shells** essential to the egg formation are the widely known common and cheap kind.

The part which water plays in duck raising is an important one. It adds volume to the food and it makes the food cost less because it makes the same amount of food more filling. Ducklings are given a vast amount of water to drink between meals, also, and this fattens them. Water fattens ducks more than any other poultry — they absorb a lot of it in the course of a day.

Does the reader know of anything raised on such a cheap and simple ration as ducks, as above described? No expensive grains are needed.

That is where the profits come in. The producing cost is low but the selling price is high. Even when sold to middlemen, the price is from **three to five times the cost of the food.**

FIRST NURSERY YARDS

The little ones are seen poking in the dirt and running around enjoying themselves. Wire netting eighteen inches high separates each pen. The birds do not fly at any age, nor jump from one pen to another over the netting.

CHAPTER TWO

SHELTER, VENTILATION

No buildings necessary to make a start with ducks on the average place.—Progress in the business means buildings. —Simple structures, all with dirt, not board, floors.— Kind of buildings needed depends on where you live, whether in warm South or cold North.—Run the first incubator in cellar of house or back room.—Dimensions of lumber for house construction.—Pens in the nursery should not be more than three feet wide.—Some things about ventilation important to remember.

Until one gets a large plant in operation, buildings for ducks are a secondary matter. The average home place with a little land is big enough to make a start. Quite a business in ducks can be done on limited ground. (See our free booklet for one instance.)

We know of duck plants where double the number of breeders are kept to produce less than half the number of ducklings we market yearly. We produce from sixty to seventy-five ducklings to each breeding duck a year, while the breeders above referred to do only about thirty. We relate this as proof of what a strain of Pekins will do **when it has been built up by selection, and attention to the details of the breeding. It is a waste of energy, time and money to keep fifteen hundred birds producing eggs when half that number will do as well.**

Every house has a cellar or back room where the first incubator may be run. The few breeding ducks may be housed in the woodshed or small building or shack of any kind. Not even wire netting eighteen inches or two feet wide is needed to confine them; boards fifteen inches wide will serve. Use carriage house, barn or outhouse. The mature breeders can stand any amount of exposure in our

EXTERIOR OF BROODER HOUSE

The hot water heater (coal for fuel) is in that end of the house nearest the eye. The brick chimney leads straight up from the heater pit. Note the ventilators at regular intervals in the roof. Never build a tight house for ducks but always provide for ventilation.

The small structure in the foreground is a simple shelter for ducks outdoors to protect them from the sun and the rain. The posts are three feet high. The ducks know enough to go under the shade without urging, whenever their well-being demands that they should.

SHELTER, VENTILATION

winter, but they should have the chance of getting in under where it is dry, and where they can squat on dry leaves or other bedding so as to keep their feet warm. If a freezing night comes and you have your breeding ducks in a very cold shelter, better get them into the barn or other fairly warm place where their eggs will not freeze.

After April, in the North, they can lay anywhere safely without danger of frozen eggs.

If you are in the South, or any State where the climate is warmer than ours, you should handle your ducks, as far as shelter is concerned, as you observe poultry raisers do whom you know. Understand, the pictures and descriptions of buildings which you see in this book apply to cold New England. Duck breeders here put up expensive, substantial buildings, some with hot-water heaters, burning coal, and the fact that they can do this, covering their farms with such structures, is proof of a substantial kind that there is money in the duck business.

When the youngest ducklings come out of your incubator, they need a brooder, or foster mother wooden device. If there is anybody who reads this book who does not know what a brooder is, the picture of it will tell, and the machine, with the directions that go with it, will be understood at once. Brooders are used both indoors and outdoors. An outdoor brooder, however, should not be put out in very cold weather, just because it is labelled for outdoors. Protect it all you can in such weather by putting it in a shed or under cover somewhere.

The little ones are managed in a small portable brooder in the same manner as described in the chapter of this book headed "Youngest Ducklings." What is written there applies to your little ones, only we describe them there as having a big house over their heads, whereas your brooder is a small house in itself.

Progress in the duck business means buildings. There are single brooder houses, double brooder houses, cold houses, fattening sheds, incubator cellar, killing and shipping house, grain storehouse, and so on.

INTERIOR OF BROODER HOUSE

This is the nursery for youngest ducklings. The pens are three feet wide. Cross boards are set halfway in the pens, as pictured, so that the little ones will not wander far from the hover. The board tops of the hovers are seen. (The milk can is on top of one section.) The hot water pipes are directly underneath the board tops. The pipes (bent) are for the purpose of carrying water from a central pressure supply so as to save labor when filling the small drinking fountains used for the youngest birds.

The timbering of the brooder house is well illustrated in this picture. As shown, the roof is double uneven, with the long side facing the sunny, or south side, and the narrow roof facing the north. A house of this construction is made any length, to suit the number of ducklings which it is desired to handle.

SHELTER, VENTILATION

Ducks should always be on the ground. Do not have floors in any duck houses.

The ordinary brooder house is built with an uneven double roof (not single roof). That is, the back roof is half as long as the front roof. It is high at the back to give walking space for the attendant; this form of construction gives headroom there. Erect the house so that the long side of roof will face the sun, that is, the south.

For a house under one hundred and fifty feet long (sixteen feet wide) use lumber of the following dimensions: Studs two by three inches, plates two by four, sills two by six, rafters two by five, collar beams one by six. For a house thirty feet wide and over one hundred and fifty feet long, use studs two by four, sills three by five, rafters two by six, collar beams one by six, and for plates two two-by-fours spiked together. In a house thirty feet wide or over use collar beams two by five. A house of this width should have posts to hold up the roof.

Lay the sills of all houses on posts, or brick or stone piers. Set the piers about five feet apart.

In houses built with even double roof, the walk is down the middle, under the ridgepole, and not down the back. Such a house has pens on each side of the walk.

Good, substantial duck buildings can be erected cheaply provided that roofing paper be used instead of shingles. **There is a great difference in roofing papers, however.** Many have to be painted frequently in order to keep them efficient, and **the cost of this paint, with labor of applying,** will soon amount to more than if shingles had been used at first.

The pens in the nursery house should not be more than three feet wide. Some recommend that they be four, five or even six feet wide. Not more than fifty little ducklings should be put into a pen, and fifty will go into a three-foot pen all right. Those who have built wider pens have found it not wise to put more than fifty into the pen. They will crowd together anyway, and more than fifty in a bunch may make trouble by walking over one another. The stronger ones will tramp over the weaker ones and hurt them. No lanterns are used in the nursery at night to keep the ducklings from crowding, because the little ones are under the brooder covers, shut in the darkness, where the lantern light could not penetrate anyway.

In a house with pens only three feet wide, of course the ducklings should not be kept longer than three weeks. After that age they should be transferred to more roomy quarters.

Ventilation. Ducks need fresh air, and in building the houses provision should be made for ventilation. **Tight houses are not to be built.**

If you have a tight house full of ducks, and keep them there three or four days and nights in bad weather, the inside of the house will grow very warm, and ammonia from the manure will rise, making your own eyes and the eyes of the ducks smart.

Every one-hundred-foot house should have two ventilators through the roof, and also should have windows at the back to be opened when needed. These windows should be managed according to the weather.

Early in the morning, when you go into a house filled with ducks, you will see the hot air and ammonia fumes going off from the ventilators.

If no provision whatever is made for ventilation, and a large number of ducks are kept in a house, the air will get so bad that some ducks actually will go blind from the irritation of their eyes by the ammonia rising from the manure.

When the weather in the spring begins to get warmer so that there is no danger of the eggs freezing, take out the windows entirely, so that the air will circulate freely from that time on, all through the house. Nail laths or wire netting over the windows to prevent the ducks from getting out at night, and also to prevent cats and other animals from getting in. You must manage the windows so that the eggs will not freeze. Be governed by the time of year and the weather.

CHAPTER THREE

CARE OF BREEDING STOCK

Twenty-five ducks and five drakes should be put in each pen.—Pens separated by boards inside the house and by wire netting outdoors.—Growing food is the same they had while at killing age.—Fed twice a day and given plenty of water.—Whole corn in the ration to strengthen the layers.—Bedding the pens for winter.—How to prepare the vegetable food.—Ducks are fond of worms, and root after them greedily.—Lanterns necessary at night.—Ducks enjoy snow.—Breeders lay better in winter if let outdoors daily.—Price tells whether stock offered for sale for breeding is young and good or old and played out.

Ducks and drakes which we sell you for breeding, and ducks and drakes which after the first year you save for breeding stock, should be handled as this chapter advises. If in winter, house them.

Thirty head should be put in each pen: twenty-five ducks and five drakes. Allow ten square feet for each bird. That is to say, the thirty birds should have a space of three hundred square feet. A pen containing three hundred square feet would be twenty feet by fifteen feet in size, or ten feet by thirty feet. This is the space inside the house. Each outside yard for a pen of that inside size should have from thirteen hundred to two thousand square feet. If the pen is fifteen feet wide then outside the house it should be close to one hundred and thirty feet long. An outside yard ten feet wide and twenty feet long is not long enough. Yards for thirty head should be fifteen feet wide and ninety feet long.

The pens are separated from each other by wire netting which should be two feet wide. Eighteen inches is not

ON GRASS

When the ducklings are crowded out of the nurseries by the on-coming hatchings, they are put out on the range. If they are kept for breeders, they are allowed to stay on the grass. Ducklings to be killed, however, are taken off the grass two weeks before killing time and placed on dirt. If allowed to stay on the grass, their flesh will be yellow, not white, as it should be, and as the markets demand.

On meadow land, ducks of the size and age shown above have a fine time grubbing for worms. They do not scratch the surface like a hen, but get down under the dirt with their bills and grub. They eat worms greedily, as many as they can stuff into themselves all day long, until the food passage is distended to large size. This diet of worms cuts down the regular feeding ration and has a good effect on the breeding qualities of the stock, not only in producing more eggs, but influencing the fertility to a remarkable degree.

CARE OF BREEDING STOCK

wide enough for these big ducks, especially when snow covers the ground. The proper size of wire netting is No. 19, two-inch mesh or three-inch mesh. The next finer size of wire, No. 20, is not stiff enough.

In building these outside pens with the 24-inch wire netting, **drive the wood stakes first,** then tack the wire netting on with galvanized iron staples. Do not drive the staples clear home. Drive them down three-quarters of the way so that when you wish you can pull out these staples with a cotton hook. A very handy tool in handling the wire netting and staples is a hammer and staple puller combined. **Do not nail the wire netting tight to the stakes and later pull up stakes and wire netting together and roll them into a bale.** You will find this awkward and clumsy work. A roll of wire netting and stakes ninety feet long is hard to manage. The staples should be pulled out and saved and the netting rolled up separately. The stakes should be loosened with a sledge hammer and pulled up and out of the way. This is done before you plough up or spade up the yards previous to sowing them down to rye or other green stuff.

A good time in northern latitudes to plough up the yards is in August and the rye is then sown. In many places in the West winter wheat should be sown instead of rye. In the South wheat should be sown. Wheat should also be sown in California. In this matter of sowing green stuff you will be guided by what you see around you in your State.

Inside this house where the breeders are kept use **board partitions two feet high,** not wire partitions. These board partitions will prevent the wind from blowing in draughts.

Either **wooden troughs** or **wooden pails** may be used for giving the birds water in and out of this house. A **galvanized iron pail** specially made so as to have a wide base is the best. The ducks would tip over the ordinary pail whose base is narrower than its top. The ducks do not climb into the pails. They do not try to take a bath.

HOW WE MAKE DUCKS PAY

They use them only to drink from. Always when giving them water, provide a vessel deep enough so that the water **will reach above the nostrils** and give the birds an opportunity to clean out their nostrils in the water.

Two food boards are enough for a pen of thirty head, each board being five feet long and twelve inches wide, with a three-inch strip nailed around the edges. Two gallons of water should be given at a time for a pen holding thirty birds.

Ducks saved for breeders are fed on the same food which they had while growing to killing age, but instead of being on a dirt range they should be on a **grass range.**

This food is given to them **twice a day,** morning and evening. They should be watered more often during the day. This water is an important factor in their growing. They can be watered **five times a day** in addition to the two times at which they feed. The value of water is strikingly seen in the case of strawberries. By giving plenty of water to his berries the strawberry grower can double the size of his berries.

This food for the birds saved for breeders is as follows (by measure): **Vegetables, ten parts; green stuff** (or cut dried clover), **ten parts; beef scraps, ten parts; low-grade flour, twenty parts; bran, twenty parts; corn-meal, thirty parts.** In addition, **grit** and **shells** and **a pinch of salt** should be put in. About one per cent of grit is enough, also one per cent of shells. For salt, use the common fine table salt. Do not use the coarse salt such as is used in making ice-cream. That is too coarse. Nearly all animals need salt to keep them in a healthy condition. It will improve their appetites and keep their blood in good order.

About once a week put into the ration **one per cent of ground charcoal.** The object of this charcoal is to sweeten and clean the stomach of the bird. It corrects any acidity in the stomach.

The food ingredients above are **mixed up dry** first so as to get them thoroughly stirred up, then water should be

CARE OF BREEDING STOCK

added until the mixture becomes **lumpy but not sloppy.** Test it by picking up a handful. It should be lumpy and crumble but **should not stick to the hands.** If it sticks to the hands, it has too much water in it.

Use one of your food-pails to measure the various ingredients.

Feed the mixture as soon as you have it mixed. In the winter time, to save labor in the morning, you can mix at night and have it all ready to feed in the morning.

A pen of thirty ducks should eat **about a pailful** of this at each feeding.

When ducks are laying well, they should be given a **quart of whole corn** at noon. We mean that each pen of thirty ducks should have a quart of whole corn. They like it and begin to ask for it about noontime, each day. The object of this whole corn is to stimulate and fortify the system. Cut open a duck's egg after it has been germinating two weeks and you see the large amount of blood in it. Draining this out of the duck at the rate of one egg a day is quite a strain on her system and the corn is fed especially to strengthen the bird.

Clean food is essential. Do not let the food boards get dirty. When the ducks are through feeding turn the food boards over, face down, so as to keep them clean.

It is said that you can get hens and pigeons too fat to lay, but this is not true of ducks. By overfeeding a hen you can stop her laying but **you cannot overfeed a duck so as to stop her laying.**

The pens inside the cold house should be **bedded** when necessary. For bedding, hay, straw, sawdust, meadow hay, leaves or **baled shavings** can be used. The **shavings** are best. They are cheaper and more convenient to handle. A bale of them weighing from ninety to one hundred pounds costs only about twenty-five cents.

This bedding should be put down inside the pens in the house to a depth of **one inch** to start. The foundation is dry sand or gravel. A new layer of bedding should be put down twice a week on top of the old bedding. It is

FOOD-MIXING MACHINE

This is really a machine used by bakers for mixing dough, but it is just what is wanted for mixing the ration for ducks. This size cost three hundred and twenty dollars, but it saves the labor of two men. There is a smaller size sold at half that price. The mixer shown in above picture is run by power, a gasoline engine. The steel hopper is stationary and the mixing is done by two horizontal revolving metal paddles, moving in opposite directions.

CARE OF BREEDING STOCK

not necessary to clean out the old shavings. If the **pens are cleaned out with a fork twice a year,** that is enough.

The bedding **should be changed** on some fair day when the ducks are outdoors out of the way. Do not change the bedding while they are inside of the house, for if you do it will frighten them.

The ducks tread down the manure and shavings into a hard layer. The peculiarity of this manure (for the mass is nearly all manure) is that **it does not heat and ferment in the pens indoors** but when you throw it outdoors in a pile it does heat and ferment. It is a splendid dressing for lawns or for general use about the farm, same as any manure. It is very rich and vegetation to which it is applied will thrive luxuriantly.

Grow vegetables in the summer to feed to the ducks in the winter. Anything in the vegetable line is good, such as **turnips, carrots, mangels, cabbages, small potatoes and beets.** A change in the food can be made with advantage every other day, giving the same vegetables to the birds only three times a week. For a large number of ducks, a great labor-saving machine is a vegetable cutter. Run the vegetables through the cutter so that when they come out they will be in pieces about three-eighths or one-half an inch square. The ordinary hand cutter is sufficient even for a big lot of ducks. It is not necessary to run the cutter by power. **After the vegetables have been cut, boil them in kettle, tank or caldron.** Small potatoes should be boiled and then mashed while hot with a pestle, skins and all. The vegetables should be boiled until they are soft. It takes carrots about two hours boiling to soften.

Cut clover for ducks costs about $1.50 per one hundred pounds. It is cut clover which has been cured dry.

Ducks use their bills in **rooting** as pigs root with their snouts. Loam frequented by a flock will get to look as if a harrow had been run over it. **Ducks root to get grubs and worms,** which they love and which do them good, **making their eggs more fertile.** Hens scratch the surface of the ground only, but ducks get under the surface. A newly ploughed field or a swamp is much enjoyed by ducks. They will root there until they are **filled clear to the neck with worms.** Insect life of all kinds is relished. If a young duckling by chance eats a bee or a hornet, the duckling will be injured or perhaps killed. There are many homes with orchards, sink-spout or a little meadow where a flock of ducks would be very much at home and have a fine time. It is good business to turn ducks into such places because **the worms cut down the grain bill** and are just so much clear gain in cost.

The proper number of ducks (which you are saving for breeding) **to be kept out doors in the summer in one flock is one hundred.** Do not keep more than one hundred on one grass range. If you do, they are more liable to run over each other in case of fright or panic and hurt their wings.

The average grass range for one hundred ducks should be at least fifty feet long and thirty feet wide.

The ducks are up and about all night long more or less. They rest in the sun in the daytime, but at night they do not care to be absolutely tranquil.

During a black night with no moon they are liable to take alarm and crowd over one another in fright in a corner of the pen and this will result in injured wings and lameness of some of the birds. To avoid this a **lantern** should be lighted and hung from a limb of a tree in the grass range or from a wooden post planted in the middle of the grass range. The light from this lantern will enable the birds to see the pen and one another and they do not take alarm nor crowd. The oil in the lantern lasts easily all night long. The lantern should be filled with oil every day. This item of kerosene oil for the lanterns is a small one even on big plants.

Some breeders with fancy houses and pens, who have electric light on their premises, also have electric light for their ducks, having one small twelve or sixteen candle-power bulb for each pen.

CARE OF BREEDING STOCK

Ducks in a new home or even ducks in their old home do not lay eggs together all at once. Given a certain flock of ducks, if half of them are laying by February first, **that is doing very well.** One month later all of them may be laying. We are speaking of our latitude now, remember. In southern latitudes like that of Virginia, for example, eggs are laid differently. There birds hatched in March go to laying in the fall. This is true all through the South and on the Pacific coast.

Our ducks do well in Canada. Unlike hens, they are fond of snow. They will wade around in the snow and slush and **enjoy getting out on snow** during the daytime, even in the coldest weather. However, to get eggs from them, their feet should be kept warm and this means that in cold, snowy places like Canada they should not be allowed to stand on the snow all the time but should be protected by being given a chance to go inside a house.

The breeding ducks will lay better if you **let them outdoors** every day during the winter, except when it is bitterly cold. If there is water in the yards which has not frozen, or which had been melted by the sun, they will play in it and **it will not hurt them.** After a fall of snow, it is well to shovel about ten feet from the house out into the pens so that they will have a space to rest in during the day without being liable to cold feet. They will play on the snow but they will not lay so well when their feet get cold. Of course they know enough to court the bare ground in preference to the snow, and you do not spend any time driving them from the snow to the dirt.

We believe in keeping the old birds breeding and laying for **two full years,** or until they are twenty-eight to thirty months old. Then they can be sold off for market. If sold alive, they will bring from eight to nine cents a pound —if dressed, from eleven to twelve cents a pound. They are not worth so much as a duckling because they are old and tough. They can be told by their extra large size, the darker color of the bill, and the hardness of the breastbone. The breastbone of a duckling at killing age is not hard, but soft—it is only gristle. We do not keep breeding ducks at work longer than two years because they have done a lot of work by the end of that period and are not equal to any more. At any rate, with plenty of young stock coming along all the time, there is no need of taking the risk of infertile eggs. You can always find live ducks of breeding age advertised for sale, but **the price tells** whether they are young stock suitable for breeding, or the old, wornout birds, good only for the pot. **If the price is one to two dollars each, you can be sure that the birds are not good for breeding, but are castaways, no longer wanted by their owner, who is trying to get rid of them alive for something above the market price for killed old birds.** Remember, a bird of breeding age is two or three months older than a duckling of the killing age. No breeder can afford to feed and keep his choice birds that length of time, culling out unfavorable specimens meantime, and sell them at low prices, if he has any reputation for breeding, and pays his bills. If he wanted the one dollar selling price with the least cost price to himself, he would have sold the duckling to market at killing age. The fact that he has kept the bird for months beyond the killing age and then offers it for one or even two dollars shows that he is letting it go at **less than cost**, and therefore he is getting rid of it because he has no further use for it. It is not right for anybody to sell such stock alive, representing that it is good for breeding. **The market for prime young breeding stock always is steady at good prices.**

CHAPTER FOUR

SEX, PAIRING, BREEDING

How to tell the duck from the drake.—Signs of pairing.—Birds are giants sexually.—How to corner the bird you wish, and how to pick it up.—Never walk through a flock of ducks.—Definition of a yearling.—Selection of breeders.—Three-legged and four-legged birds and other freaks.—Do not do any inbreeding.—Stock should be outbred, choosing the best specimens from each pen.—When new blood is needed.—Do not start with scrub stock, for if you do it will take you years to accomplish what you can have at the beginning by more judicious buying.

The duck is distinguished from the drake both by sight and by hearing.

The drake, when full-feathered perfectly, has in his tail-feathers **one feather which curls up** as shown in the picture. This is not an invariable test because sometimes the curl feather may have been pulled out, or lost out from moulting or other cause. The curl feather shows on the drake when he is four months old and it is a sign of puberty; in other words, that he is ready for breeding.

A duck quacks, but a drake does not. The drake makes a noise — it is a sort of low, rasping, hissing-like noise hard to describe, but **much different from the quack which the duck makes,** and which you can instantly distinguish.

A drake is generally larger than the duck, but not always.

The best and quickest way to distinguish the sex of a bird is to **pick it up by the neck** and listen to the noise which it immediately makes. When we are examining a pen of them rapidly, we never take time to look for the

DRAKE, SHOWING CURL FEATHER

This photograph of a drake shows the curl feather in the tail which is first seen at puberty and is the mark of the sex. The ducks do not have it. Sometimes a drake will be found without this curl feather, which may have been lost in moulting, or have been pulled out. In examining a lot of birds, the quickest way to tell the sex is to pick them up one by one by the neck and listen to the noise or note which each makes. See page 39 for further particulars.

curled tail feather. We pick each up by the neck, one after the other, and listen to the note, or noise, made by each, the decision being made instantly.

Fertilization Necessary. The act of copulation (connection) of the male and female is a plain, unmistakable ceremony, same as in the case of rooster and hen, and not so blind as in the case of the smaller birds. We think we ought to discuss this subject because it is a very important one, the foundation of the business. Ducks will lay eggs, but not fertile eggs, unless they have had intercourse with a drake. We have known a beginner to run an incubator the specified time filled with eggs laid by ducks with which no drake was allowed to run; result, no ducklings from said eggs, which was quite a puzzle to the beginner until she thought the matter over. Some people have no imagination, or take everything for granted, or are actually ignorant about the simplest things. Given a duck's egg, it does not follow that the egg is fertile. **The beginner should be sure that the eggs which he puts into the incubator come from ducks which have received the attentions of drakes.** It is not uncommon for incubator experimenters to fill up the machine with eggs bought at a grocery store. These beginners do not know, or seem to care, whether eggs have been fertilized or not.

The duck in love-making makes the first move by bowing her head low, then the drake bows his head low. They bow faster and faster; the duck squats and permits the drake to mount her. He seizes her with his bill by the back of her head, wearing the feathers off as the season progresses. Connection being quickly performed, the drake drops, or falls, from the duck, and bolts away from her five or six feet, then stops.

When you see the ducks and drakes bowing to each other, this is the sign that they wish to pair, and will, then or soon.

The first active period of heat comes on when the duck or drake is about five months old. In excessive heat, and without drakes, ducks sometimes will mount each other,

HOW WE MAKE DUCKS PAY

acting like two cows, or two female rabbits. **Do not be led astray in the matter of sex by observing such actions.**

One service of drake to duck will fertilize a dozen eggs. One drake will attend five ducks. **Twenty-five ducks to five drakes** is the most successful proportion, no more.

Drakes have no preference for a certain duck in their love affairs. As a rule, they do not quarrel among themselves over a duck or ducks, or interfere with one another. Sometimes all the drakes in a pen except one will annoy or persecute that one. He seems to make all the rest jealous or irritable unanimously, and they unite to prevent him from paying attention to any duck in the pen. Such a drake **should be removed from the pen,** as the others will make his life miserable.

The drakes are strong, and one can force a duck to have connection against her will. This is not true of most animals. It is best not to interfere with him. At first, the ducks will invite him; as the period goes on, they will not. The drakes act in a bold, aggressive manner all the time except when they do not wish to pair. Then they go about with their heads and necks drawn in. When they are making love, they extend their heads and necks, and walk proudly.

The drakes keep cleaner and whiter than the ducks. Their bills are a darker orange color.

Drakes and ducks are stronger sexually than other animals their size, or anywhere near their size, and their reproductive work proves it. The testicles of the drake (inside the body) are of extraordinary size, as large as eggs or lemons. The duck lays an astonishing number of the big, fertile eggs.

How to Catch Ducks. Suppose you wish to pick out certain ducks or drakes from a pen, for breeding or for killing, or for examination, or for any purpose. **Do not go at it single-handed.** Get a helper. Also a board ten feet long and a foot wide. Give your helper one end of the board and you take the other end. Both you and your helper then walk to a corner of the pen, holding the board

SEX, PAIRING, BREEDING

between you, driving the ducks along ahead of the board into the corner. Then you can reach the duck or drake you wish and pick her or him up by the neck. **Never attempt to drive more than twenty at a time into a corner.**

How to Pick Up Birds. (This does not apply to youngsters, but only to the grown ducks and drakes.) Always catch a bird **by the neck,** in a firm grasp. Never pick up a bird by the body. If you do you will make trouble for yourself and the duck. You are liable to bend, or twist, or dislocate or break the bones of the wings and legs. You can carry three or four ducks, or more, as many as you can lift, between two hands, in front of your body. **Do not be afraid of strangling the ducks or breaking their necks.**

The above applies only to birds beyond the age of four weeks. Little ones just out of the incubator can be gathered by armfuls or apronfuls. From one to four weeks of age they can be handled best by picking them up with the hand around the body, and this will not hurt them.

Walk Around, not Through. Never walk through a flock of ducks, young or old. Always go around them. An inexperienced person always will try to go through a flock, which will confuse the ducks, and some will crowd under his feet and will get hurt either by his feet or by themselves as they jostle one another. Just walk slowly and take your time and they will get by you. Do not rush around among them.

What a Yearling is. A yearling duck is not a duck that has lived a year, but **a duck which has passed through one year's work.** For example, a duck may be hatched in March, but she will not be a yearling duck the following March. She will not be a yearling duck until the end of the second December following the March in which she was hatched. By that time she has **passed through one year's laying.**

Selecting Breeders. The selection of breeders should be going on all the time while you are killing. **Select the**

FIGURE 1.   FIGURE 2.

HOW TO HANDLE LIVE BIRDS

This is the way to pick up and carry the live ducks and drakes which are four weeks old or more. (Do not handle the little ducklings just out of the incubator in this manner.) Catch the birds by the neck in a firm grasp, as shown in Figure 1. Never pick up a bird by the body. If you do you will make trouble both for yourself and the duck. Carry them from pen to pen, or from pen to killing-house, as shown in Figure 2. Do not fear that you will strangle the ducks or break their necks.

You can carry three or four birds, or more, as many as you can lift, between two hands, in front of your body.

SEX, PAIRING, BREEDING

very best and save them for breeders and kill the balance for market. To examine a bird in order to find out whether it should be saved for breeding, pick it up by the neck, look at it and feel of it.

In selecting breeders, first look for the largest, but do not take a large bird that is sluggish in behavior. The big, active ducks make the best breeders. Do not take any birds which have black feathers. Black feathers are not necessarily a sign of bad blood. They may be what is known as a " throw-back " in the breeding or a " sport," liable to come out in any breeding, no matter how pure the stock is.

In breeding large numbers of ducks there will occasionally be freaks, such as three-legged and four-legged ducks. We have seen three-legged and four-legged ducks grow until we killed them for market. The supplementary legs were not as large as the two legs on which the duck walked, but were dwarfed.

Sometimes a duck will be found with more than one egg orifice. We have seen freak ducklings with as many as five egg orifices.

No Inbreeding. We do not do any inbreeding. Our stock is cross-bred. In the years that we have been building up our strain, **we have greatly increased the size of the breast and the depth of the keel.** By poor selection of breeders from year to year and thoughtless inbreeding, some breeders of Pekins of our acquaintance have weakened their stock, and cannot now produce the bird they ought.

The breeder should aim to introduce new blood into a certain pen or pens every other year, to keep up the size and vitality. A drake which, for example, has served three to five ducks for two seasons should be replaced by another at the end of that time.

We do not do any pedigree or record-keeping. Such drudgery is a waste of time and is a wholly unnecessary detail. By separating the birds into flocks, each with its drake or drakes, there cannot be inbreeding.

HOW WE MAKE DUCKS PAY

It is surprising what improvements a duck breeder, if he is careful and intelligent, can make in his flock in size and good laying qualities, by **selection of the best specimens** from year to year. We have got our birds up to the point where they produce **twice the number of fertile eggs** which the earlier generations did years ago, and the birds have a longer and deeper keel, and weigh more. These are important factors in money making, especially the matter of fertility. Those who start with our breeding stock get the benefit of the years of study and selection we have done. They should continue the work with their own breeding stock, remembering to save for breeders only the largest and best birds.

Do not get the idea that you can buy any kind of a white Pekin duck and in quick time build up a strain of superior size and breeding qualities.. You might just as well try to breed a race-horse out of common drudge horses. The blood that makes for size and fertility can be produced only by years of effort. **It is much better to start with a trio of stock at ten dollars a head than to buy a bargain lot at two or three dollars apiece,** or even less, as they are sometimes sold. The precious seed eggs from first-class stock only are what you wish to secure from the ducks you are breeding. **By buying our breeding stock you jump over years of effort which you will have to go through if you start with scrub ducks picked up at bird and poultry stores,** or bought here and there of anybody who happens to have a few ducks.

CHAPTER FIVE

THE EGG

Laid at night by the breeding birds.—Gathered in basket in the indoor pens.—Washed in cold water.—Handled occasionally so that they will not spoil.—Selected for the incubator.—Turned twice daily in incubator to give new albumen to the germ.—Tested by lamplight.—They pip on the twenty-fourth day.—Hatch should be complete on the twenty-eighth day.—Leave the ducklings in the incubator until they dry, then take them to the brooder.

Shut up the breeding ducks and drakes at night and do not let them out until eight o'clock the next morning. **They lay their eggs during the night and early morning.** By eight o'clock in the morning they have all laid.

Fill the water pails in the outside yard and let the ducks out to drink, then go from pen to pen inside the house and **gather up the eggs.**

Sometimes a very cold night will come on in our latitude. In that case it is necessary to go around as soon as you can in the morning to get the eggs, so that they will not have a chance to freeze. (This applies only to a **cold house.** In a warm house you would not have to watch out for frozen eggs so carefully.) One frosty morning, in the value of eggs spoiled by freezing (provided you are not on guard), would more than pay for the **heater** necessary to warm the house.

If the weather is stormy outside, you do not let the ducks out of the house, but go around from pen to pen among them slowly gathering up the eggs. Do not go hurriedly, for if you do you will scare them.

Use a basket to gather up the eggs. Some eggs are in the shavings where the ducks have hidden them. Sometimes

WHERE THEY LAY THEIR EGGS

This house is heated by hot water so that the eggs will not be frozen on cold winter nights. The pens have a dirt (not board) floor and shavings are put down for bedding. The partitions between the pens are not made of wire but of boards, so that the pens will not be draughty. The ducks lay at night. Some try to hide their eggs in the shavings but they are easily found, and the same ducks will try to hide them in the same places the following night. Let the birds outdoors at eight o'clock in the morning and then go from pen to pen and gather the eggs in a basket.

THE EGG

a duck will make a nest three or four inches under the shavings, then lay in the nest and cover the eggs with shavings. She will do this night after night in the same place, so after you have once located the place you can go to it morning after morning unfailingly. **Most of the eggs are in plain sight, on top of the shavings.**

The first eggs of a duck are infertile. After she has laid several eggs then they begin to run fertile. The first eggs should be used for cooking or sold at market, where they bring from **thirty cents a dozen** up. They almost invariably bring more than hens' eggs and there is a good demand for them. They are a trifle different in taste from hens' eggs. There is some prejudice against duck eggs in some markets on account of the **fishy taste** found in the eggs of common or puddle ducks. Ducks which are fed on grain lay fine eggs of good flavor. As far as cooking goes, duck eggs taste the same as hens' eggs and **a duck egg will equal two hens' eggs.**

The number of eggs laid by the duck steadily increases. By February 1, in our climate, about half of them should be laying. We begin to save eggs for the incubators as soon as we see that the production of eggs is increasing steadily from day to day.

The eggs should be taken in the gathering basket to a cellar and **washed** there, as soon as gathered, **in cold water** at a temperature of about **forty-five degrees. Do not use ice water or hot water.** If you use ice water you will freeze or chill the germs in them. If you use warm water you will start the germs to growing.

Use a cloth to wipe the eggs clean. After washing them, put them in a basket in the cellar and keep them there at a temperature of about **forty-five degrees** so that they will not freeze and so that germination will not start. The ordinary cellar of a house is just the right temperature. The air should be good in the cellar, that is to say, not foul or close, because the egg shells are porous.

While they are being kept in the baskets in the cellar, **the eggs should be handled every three or four days** so that they will not spoil. That is to say, they should be taken up and turned. **The object of this is to prevent the yolk from sticking to the sides of the eggs.** In warm weather, if the egg is allowed to stay in one position continously, the white will get soft and the yellow will go through the white to the shell.

The eggs should be kept in the cellar no longer than two weeks before putting them into the incubator. We have kept them a month, but not over two weeks is the best time.

If the temperature of the cellar rises to fifty-five or sixty degrees, a slow process of incubation goes on inside the eggs. Keep the eggs over two weeks in a cellar at sixty degrees instead of forty degrees and they will hatch earlier than others.

In selecting eggs for the incubator, **do not take all of them.** Take only the perfect eggs. Eggs should be rejected which are small, large, ill-shaped, and have holes or pit-marks. The selecting or sorting of eggs might be done after washing and the rejected eggs used for cooking. Two weeks in the cellar would not spoil the eggs for household use.

Preparing the incubator. The incubator should be cleaned and put in order, the lamp lighted and the **machine run empty for two days at a temperature of one hundred and two degrees.** Then put in the eggs. The temperature **will fall at once** because the eggs are cool, but **do not fuss with the regulator.** Leave it alone and let the temperature rise to one hundred and two degrees, as it will slowly in twenty-four hours. The thermometer then will be one hundred and two degrees again.

After the eggs have been in the machine thirty-six hours take the tray out of the machine, put it on a table nearby and **turn the eggs with fingers or hand.** Run the flat of the hand over them so as to roll them around, stir them. It is not necessary to turn them an exact halfcircle. The object of turning them is to supply new **albumen to the germ. The embryo feeds on the white of the**

THE EGG

egg. **When you turn the egg with your hand you give an opportunity for new food, new albumen to get to the embryo. A hen turns her eggs in the nest twice a day for the same reason instinctively.**

The little duckling is made wholly from the white of the egg. The albumen contains the feathers, flesh, everything. The object of the yolk is to furnish food for the duckling during the last few days of its life in the shell.

After that first turning of the eggs thirty-six hours after having being put in, the eggs should be turned **morning and evening.**

The incubator **cools off** more or less while you are turning the eggs on the tray on the table nearby, **but this should cause no alarm.** When a hen leaves her nest, as she does to go off for food, eggs which were being covered in the nest are cooled off.

Testing the eggs. On the evening of the seventh day after the eggs have been put in, they should be **tested.** That is to say, if you put the eggs in on Saturday they should be tested the following Friday. Duck eggs have a white shell which is transparent and it is **very easy to test them.** We have a box big enough to hold a lamp (see illustration). The box has a hole cut in the front of it a little smaller than the eggs. Felting is glued around the hole so that in handling the eggs rapidly they will not be knocked and broken against the wood. These egg-testing lamp-boxes are made and sold in various forms.

The operator should take four or five eggs in his hand. He can test them very rapidly, as fast as he can pass them in front of the opening. The light is confined by the hole in the box and when the egg is put over the felting the hole is stopped and all the light from the lamp must shine directly through the egg.

If the egg is a fertile one and has been germinating while it has been in the machine, you will see inside of the egg something like a spider. Veins will cover almost the entire egg. You can see the speck forming the eye of the duckling and in fact the little duckling itself. The

INCUBATOR CELLAR, SHOWING MACHINES WHICH HATCH THE EGGS INTO DUCKLINGS

THE EGG

development covers nearly the entire interior of the egg. **If you do not see this development, you will know that the egg is infertile.**

If you find an egg which is cloudy or addled, and without the spider-like network of veins, **it is not good.** Eggs that are addled can be mashed up, shells and all, and fed to your growing ducks with their grain mixture.

EGG-TESTING LAMP

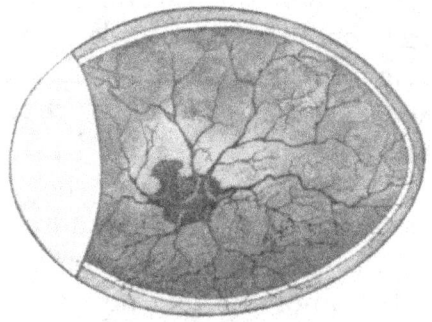

FERTILE EGG, SEVENTH DAY

If the egg is fertile and has been germinating in the incubator for seven days, you will see something like above picture.

The white space at the end is the airspace. The white line all around the egg is the membrane just inside the shell.

Any eggs which you find **black** or **bad-smelling** should be thrown away.

The **clear eggs** in which you find no germ and which are not addled **you can use in cooking,** or **you can sell them.**

This testing should be done in the incubator house alongside of the machine. In very cold weather we use a cloth and cover the eggs while we have them out of the machine so that the heat can be kept in them as much as possible. Test a whole trayful before putting them back into the incubator. Work quickly in cold weather.

There is no more testing of the eggs, but you must turn them in the incubator night and morning, twice a day, right along until hatching. If an egg shows black while you are doing this at any time, such an egg should be thrown out, for it is rotten. If when you open the incubator you detect a bad smell, you can find the egg which is at fault, if you cannot see it, by running your nose along the eggs held close to them until you come to the bad one. It should then be thrown out.

The eggs should begin to pip on the **twenty-fourth day** after being put into the machine. When you see this pipping starting among the eggs, **stop turning them** and do not open the machine until the hatch is over. The hatch should be complete if the machine is run right along on the morning of the **twenty-eighth day. Remove all eggs and shells that have not hatched at that time.**

The temperature of the machine for the first three weeks should be **one hundred and two degrees,** and for the last or fourth week **one hundred and three degrees.**

In minor matters not covered by our directions given in this chapter, **run the machine according to the directions which go with the incubator.**

After hatching, **the ducklings should be left in the incubator** from twelve to thirty-six hours, **until they have dried off.** Just before the eggs hatch, the ducklings absorb the yolk and live on that for thirty-six or forty-eight hours. **If they are taken out of the machine too quickly they are not hungry and will not eat.** You should leave them in the machine from twelve to thirty-six hours so that they will be not only dried but hungry.

Take them out of the machine gently. Pick them up or guide them with the fingers by handfuls and sweep them into the basket. Then remove them to the brooder or brooder house. The temperature of the brooder or brooder house should be near ninety degrees. The food and water should be already in the brooder before you put the little ducklings in from the incubator. They will be hungry and will go to eating and drinking at once. **See next chapter.**

CHAPTER SIX

YOUNGEST DUCKLINGS

How to handle and feed them from the time they leave the incubator until they are eight weeks old.—First food in the brooder or nursery.—When to feed it, how to feed it and how to water them.—The little birds are tame.—Weaning food begins on third day.—How to mix it and give it.—When two to three weeks old, they are taken from the nursery to the second brooder house. —When five weeks old they can go to the cold house, where, for the first time, outdoor feeding is practised.

You have allowed the ducklings to stay in the incubator twenty-four hours after they have come out of the shells. They are double the size of a chicken and in appearance about twice the size of the egg-shell which enclosed them.

A period of twenty-four hours in the machine dries them so that they are strong enough to stand on their feet. They have absorbed through the navel the yolk which surrounded them at birth and this sustains them for the twenty-four hours.

At the end of the twenty-four hours put them into a basket and carry this basket to your brooder or nursery of first brooding house. Put them in the runs, which are three feet wide and two feet deep and nine feet long. The hot-water pipes are not on the ground but are eight inches above it. On top of the water pipes is a wooden cover, one for each pen. The object of having the hot-water pipes **above the ducklings** is to give them heat from the top such as they would get from under the natural mother. Bottom heat would weaken the legs of the ducklings and is not natural. The little creatures huddle up closely to each other under the hot-water pipes so as to

DUCKLINGS TWO WEEKS OLD

They are shown here in the yards of the first nursery house. When this picture was taken, the small slides through which the birds pass from house to yard were all open. These slides are left open during the day so that the ducklings can pass in and out at will. At night the ducklings are driven into the house and the slides closed.

YOUNGEST DUCKLINGS

get the heat on their backs. Should they touch the hot-water pipes they cannot be burned.

Take these youngest ducklings into the nursery just before noon, the warmest part of the day.

Their **first food** is found by them there ready for them. There is a water drinking fountain in each pen, the **No. 1** or **smallest size** (see illustration). This water dish, like the others, is arranged so that the ducklings cannot jump into the water and get damp and also so that no water can stand in it for any length of time. A self-feeding reservoir fountain is exactly what is not wanted. The water should be renewed at each feeding time. It is not necessary to scald the fountains but they should be rinsed out. They may be scalded, say once a week.

The first food includes **bread-crumbs** and **rolled oats**. The **rolled oats** is ordinary oatmeal costing from $3.25 to $5.00 a barrel, each barrel weighing one hundred and eighty pounds.

The **bread-crumbs** are made from stale bread by running the bread through a cutting machine. Buy dry, stale bread from the bakers for about one cent a pound, twenty dollars a ton. Also use up old home bread which is dry and stale. Bread-crumbs for a small number of ducklings can be prepared from the bread by hand without a machine.

Take half rolled oats and half bread-crumbs to make this first mixture. Take them by measure not by weight. Use for a measure an ordinary wooden pail if you please. Take one pail of rolled oats and one pail of bread-crumbs or two pails of rolled oats and two pails of bread-crumbs, and so on. **Put in five per cent of good, sharp, ordinary sand with the bread-crumbs and rolled oats.** The object of this sand is to provide grit, which the little ducklings need as well as the old ones. **Two handfuls of sand to each pailful of mixture is what we mean by five per cent. Put these three ingredients, rolled oats, bread-crumbs and sand, into a box and mix them in the box.** Then moisten this mixture with water, not enough to make the mixture sloppy, **but just enough to moisten the particles.** If you

No. 1   No. 2   No. 3

WATER FOUNTAINS FOR DUCKLINGS

These three sizes must be used for successful results. The smallest, or No. 1 size, is used for the youngest ducklings. The No. 2 size is used next. When the ducklings are four weeks old, the No. 3 size is used. For birds near killing age, or over, use an ordinary pail or water trough (see below).

The No. 3 size fountain above is too large for the little ducklings. They would get inside of it and drown. The smaller sizes are not large enough for the ducklings after they have grown, because the water then would not be deep enough to reach their nostrils. The ducklings need water deep enough to souse their bills wholly in it, so that they can wash from their nostrils any sawdust or food which may lodge there.

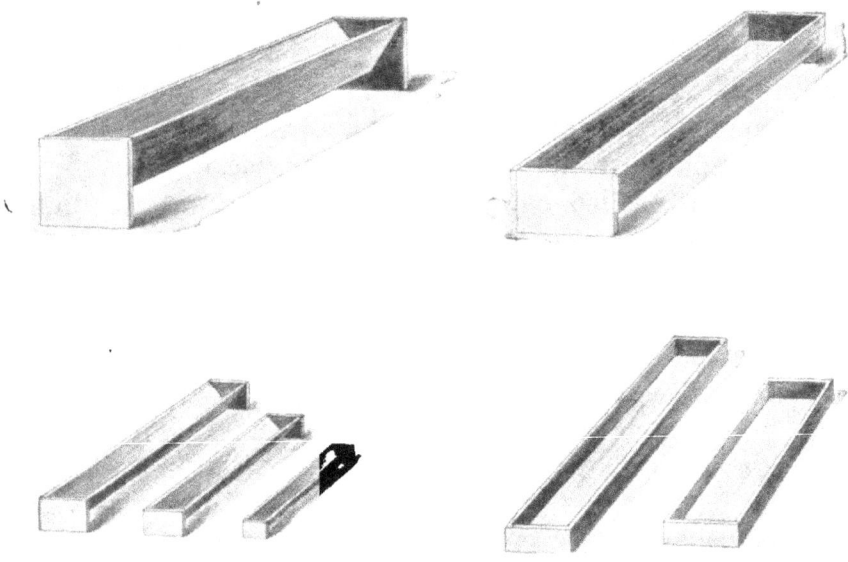

FOOD AND WATER TROUGHS

These should be built in different sizes. They are used both indoors and out as directed. There is no hard and fast rule for their size and construction. The caretaker should use them according to the age of the ducklings.

YOUNGEST DUCKLINGS

have milk you can use milk instead of water because the ducklings will grow faster when milk is used than when water is used. Understand, do not make this first mixture sloppy. Make it damp, that is the idea.

The above is the food for these newly hatched ducklings for forty-eight hours after being put into the nursery. The food is there in each pen as the ducklings are put into each pen from the basket from the incubator.

Do not stand and watch these little ducklings. They will not feed until you go away and leave them alone. The food is scattered on a board and the ducklings walk out from under the hot-water pipes to the board and eat, now and then going to the water fountain for a drink. The run-way is partitioned off half-way with a board placed there temporarily so that they will not wander too far from the hot-water pipes and get down to the window where it is cold. At night they are shut in completely under the hot-water pipes by taking this board and moving it up to the head of the pen, next the top board above the hot-water pipes.

For the first two days, the food above described should be before these youngest ducklings continuously. For that reason, visit the nursery five times a day for these first two days to renew the food on the boards in the pens and to renew the water in the fountains.

The **food board** is three feet long, the same width as the pen, and six inches wide. This has laths nailed on the ends and sides to prevent the food from sliding off or from being pushed off.

Care of Brooders. The beginner should be constantly impressed with the importance of **keeping the brooders scrupulously clean.** Every other day the droppings and dirty sawdust should be removed from under the pipes. The best way to go at this job is from the walk side of the brooder. Take off the cover. With a narrow shingle, scrape out underneath the pipes, taking only the wet and dirty sawdust out, and putting it into a bushel basket which is carried outside when full; or, in a larger brooder house,

THOUSANDS OF YOUNGSTERS

This shows how they look on a pleasant day out in the yards of one of the nursery houses. The narrow strips of wire netting which separate the pens cannot be seen in the picture, but they are there. Ducklings this age are quite an attractive sight. The small sliding doors are always up when the birds are out. They were dropped when the picture was taken so as to get all the ducklings in.

Each pen of the youngsters is handled precisely the same as the next. In the fall, the wire netting is rolled up, the wood stakes pulled up and the ground ploughed and sown (in our latitude) to winter rye. This sweetens the ground as well as furnishing green food. If the birds were allowed to run on the same dirt year after year, and make manure upon it, the ground would become tainted, affecting both the size and fecundity of the stock.

YOUNGEST DUCKLINGS

use a wheelbarrow. As each brooder is cleaned, put in a thin layer of fresh, dry sawdust from a basket taken on your arm from pen to pen.

The **food boards** should be scraped with a shingle or piece of tin each day to keep them respectably clean. Take up each board and scrape it into a basket. It will be covered with sawdust, refuse, etc.

In our system of care of brooder house, this scraping of the food boards and washing the fountains is done regularly every day after the two o'clock feeding. The attendant goes along each pen, picking up the fountains and food boards and placing them on the brooder covers. The boards are then scraped (with a sheet iron scraper about six inches square) into a bushel basket, this refuse being thrown on the manure pile. The fountains are then washed and rinsed, after which they are filled with water while standing on the brooder tops. On a plant of large magnitude, three men work together doing this job. While the cleaning of boards and fountains is going on, one of the men is bedding the pens, using dry pine sawdust for this purpose. When the task is completed, the men immediately begin putting down the fountains and food boards, and the ducklings then are ready to receive their next food.

The **five feeding times are as follows**: 6 a.m., 9 a.m., 11.30 a.m., 2.30 p.m., 5.30 p.m. The ducklings will get eager and hungry and **will cry for food** at each of these feeding times. They are not old enough to make a quacking noise, but peep. This peeping noise increases in volume until they are six weeks old. Then they begin to make a quack more like the old birds.

When washing out the drinking fountains, use a rag or dish-cloth and two pails of water. Wash in one pail and rinse in the other.

Be careful not to step on the little ducklings in giving the first food. They are very tame and will get all around your feet if you give them a chance.

Their appetites should be kept eager.

HOW WE MAKE DUCKS PAY

Do not put down too much food on the boards. The night feeding should be the biggest of the five because the food eaten then has to last them through the night.

Be sure and keep them eager and hungry. Do not load up the boards with the mixture so that they will overstuff themselves. Remember that for these first two days they are **learning to eat.**

In hot weather water them twice as much as when the weather is cooler. Fill the fountain often. They will drink this water up quickly, within five or ten minutes, then fill the fountain up again. **Do this watering always at each feeding.** The easiest way to get the water into the little fountains is to pour it from a milk-can, which is better than a dipper because it holds more and is handled easier. Do not fill these milk-cans from a faucet. That would take too much time. Let the faucet water run into a tub and fill the milk-cans by dipping them into the tub. This saves work.

Beginning with the third day, the food for the young ducklings changes to the weaning food. Mix the same ingredients of food as the first two days with **bran** and **corn-meal** in equal parts, by measure not by weight. That is to say, take one measure of **rolled oats**, one measure of **bread-crumbs**, one measure of **bran** and one measure of **corn-meal**, in other words **twenty-five per cent of each.**

By **bran** we mean wheat shells, also called shorts. It is the outside, flaky shell of the wheat. It costs about twenty dollars a ton in carload lots, but is cheaper in the West. It is a by-product of a flour mill.

Corn-meal is common yellow Indian corn which has been ground, not cracked. It costs here in the East about the same as bran. This weaning food is given for seven or eight days.

When the ducklings are seven or eight days old, cut out the expensive rolled oats and bread-crumbs and in their place in the mixture put low-grade flour, which costs about twenty-eight dollars a ton.

Remember, that all these mixtures are moistened with

YOUNGEST DUCKLINGS

water but not so as to be sloppy. **They should be damp.** When you take up a handful which has been mixed properly with water, **it will not stick to the hands** but will hold compactly together in a lump.

The food which begins at **seven or eight days of age** also has **green stuff** and **beef scraps**. To summarize, then, prepare the food as follows: **Equal parts of bran and corn-meal, ten per cent of low-grade flour and ten per cent of green stuff, such as green grass or rye or millet** (which has been chopped up in a cutting machine or by hand in a pail), **beef scraps five per cent, grit one per cent.** The scraps if too coarse should be screened, the fine part being used for the small birds and the coarse part for the older ones.

For **grit** use common sand and gravel off the farm for the first three days of the duckling's life. From then on **use our special grit made from granite in two sizes, fine and medium.** Use the fine grit at first and the medium grit as the duckling gets older.

Keep the ducklings in the first nursery house two to three weeks, depending on how you are fixed for room. If you have a hatch coming off from the incubator, clean out a sufficient number of pens in the nursery to make room for the new-comers.

The nursery house has a **dirt floor,** not a board floor. This dirt (sand or gravel or dry loam or clay) should be in each pen **with dry sawdust** laid down on top of the sandy bottom. Carry this sawdust into the nursery house in a wheelbarrow and shovel it from the wheelbarrow into the pens, then rake it level to a depth of an inch. Use dry pine sawdust. **Almost any sawdust except oak can be used.** Do not use oak sawdust for if you do it will turn the drinking water blue as it gets off the bills of the ducklings, and this bluish drinking water does not smell or taste wholesome. Pine or spruce sawdust is good.

In the Second Brooder House. When the ducklings are two or three weeks old, take them in a basket, a pen at a time, to the second brooder house where the pens are

DUCKLINGS IN THE SUN AT MID-DAY

These birds are seven weeks old. Youngest ducklings are harmed by the hot rays of the midsummer sun, and should be given a chance to get into the shade. Remember that ducks at all ages should have access to shade.

YOUNGEST DUCKLINGS

four feet wide instead of three feet wide, and ten feet long instead of nine feet long.

All ducklings are fed four times a day in this second house at the following hours: 6 a.m., 10 a.m., 2 p.m., 5.30 p.m. This second house has a hot-water heating arrangement exactly like the nursery house, except that the pipes are further from the floor.

The outdoor runs of this second brooder house are twenty feet long. **It depends on the weather** whether or not you let the three weeks' old ducklings outdoors into these runs from the inside pen. On bright, sunny days, not too cold (if in winter) you can let them outdoors, and their exercise outdoors will do them good. Remember, just now we are talking about our latitude and our winters. If you live in southern latitudes or in a warmer climate than ours, or if it is summer-time with you in this latitude, you can let the young ones outdoors more freely. **Do not let them out in the rain or snow.**

All the feeding in this second house is done **inside the house,** same as in the nursery house. The food boards in this second house are **larger** than in the nursery. They are four feet long and nine inches wide or just wide enough to be cleaned with a shovel. After each feeding time, scrape off the sawdust, refuse food, etc., from each board with a shovel and throw this refuse into the walk of the house. Every four or five days this refuse should be raked into a pile and carted out in a wheelbarrow.

The **No. 2 water fountains** used in this second house are made in the same style as the fountains used in the nursery, only **larger,** and hold twice as much water.

When the ducklings are four weeks old, change this No. 2 drinking fountain to the larger or No. 3 size.

You may ask why would not one size of water fountain be all right for all ages. The largest, or No. 3, fountain would be too large for the little ducklings. They would get inside of it and drown. They would also struggle to reach the water and would weaken themselves. The smallest size is not large enough for the ducklings after they

65

DOUBLE BROODER HOUSE   YARDS ON BOTH SIDES.   WALK THROUGH CENTER OF HOUSE

YOUNGEST DUCKLINGS

have grown because the water then would not be deep enough to reach their nostrils. The nostrils of a duck can be seen. They are two open holes at the base of the bill. The ducklings need water deep enough for them to souse their bills completely in it so that they can wash from their nostrils any sawdust or food which may lodge there.

To the Cold House. When the ducklings are five weeks old, they are taken (on a large plant) from the second house to a third, called a **cold house,** that is to say, it has no heating apparatus. If the weather is cold when they are five weeks old, use your own judgment as to putting them into the cold house. Wait until a warm, sunny day. It depends on the season and the locality. Ducks at this age can be driven in large flocks.

Feed and water the ducklings **outdoors** in the pens of the cold house. They do better if fed outdoors. It depends on the weather. If you put them into the pens of the outdoor house in the morning of an early spring, and a northeast storm comes up cold and raw, drive them back into the house and shut the doors and windows. **The ducklings would not know enough to go into the house away from the storm.** They would go to meet the storm, as far from the house as they could get, at the end of the run. The rain would not drive them in. The ducklings would stand up as straight as they could so as not to get their backs very wet, but they would not know enough to go into the house. The rain would beat down upon them and exhaust them and before long the little creatures would fall down exhausted and this exposure might kill them. Remember we are talking now about the young ducklings. **The old ducklings know enough to go in out of the rain.** When the storm breaks, they will scoot for the house. The ducklings eight, nine, ten and eleven weeks of age, and older, will stand rain in our latitude.

May is the worst month in New England to watch for weather in managing young ducklings.

The cold house is the home of the ducklings from five weeks until they are eight weeks old, when they are ready for fattening. The object of the cold house is simply to keep them out of the rain and snow. In the summer time or in southern latitudes, an orchard which has shade trees will do for ducklings when five weeks old except on days when big storms come up. On those days the young ducklings must be housed.

Distributing the Food. Where only forty or fifty pails of food are used daily, the best way to get it to the different houses and yards is by use of a wheelbarrow. The No. 7 size wheelbarrow will take ten or twelve pails. If water must be conveyed, milk-cans (previously mentioned) are best to use, as the water will not spill in carrying them in the wheelbarrow. If, however, several hundred pails of food are used daily, a large, four-wheeled truck may be used to advantage, providing the land is level enough to allow the men to draw it, otherwise it is best to use a horse and wagon.

WINTER SCENE. BROODER HOUSE

CHAPTER SEVEN

FATTENING

Growing to kill from eight weeks to market age.—The fattening sheds are simple roofs erected on dirt ground.—The birds are fed three times a day.—Secret of feeding properly the fattening mixture.—Keep their appetites eager.—Judge by the appearance of the duckling when it is fat enough to kill.—Look out for sour food.—How to utilize food which has soured.—How to avoid yellow skin.—Winter rye for green stuff.

The fattening starts when the ducklings are eight weeks old. They are driven from the cold house at this age into **outdoor (or fattening) sheds. These are sheds which are simply a roof on posts, the sides being open.** The roof should be tight but it is not necessary to make it absolutely tight by shingling or paper. All that is necessary is to nail battens over the cracks between the boards which form the roof. The fattening sheds are used from the end of April to November 1, in our latitude. In southern latitudes their use could begin earlier.

In the fattening sheds, feed three times a day, morning, noon and night, 6 a.m., 12 m., and 6 p.m. Feed a mixture of corn-meal, low-grade flour, beef scraps, oyster shells and grit, and green stuff when you have it. Mix the food in these proportions: Three parts of corn-meal, one part of low-grade flour, three-quarters of a part of beef scraps with about three per cent of oyster shells and grit mixed equally, then one part of green stuff, when you have it. The ducklings like the looks of the mixture better and eat more of it when the green stuff is in it to give it color.

This food is given until the ducklings are ten or eleven weeks old, when they are killed. **This is a rich food mixture to fatten.**

READY FOR FEEDING

On a large plant, wooden pails will be found to be the best means for carrying the food to the ducks. After the food is mixed in the house, it is put into the pails, which are stacked up as shown. If the ground is level and a large number of the pails are to be transported to the ducks, a four-wheeled truck (not shown in picture) drawn by two men is the best device for wheeling them. If a smaller number of the pails are to be transported, wheelbarrows are the handiest.

Wooden pails are cheaper than galvanized iron and they are better because they may be nested when empty and a lot of them carried by one man. Metal pails might be nested but their combined weight on the arms of the operator would be very tiresome.

FATTENING

The secret of feeding properly this fattening mixture three times a day is to feed **just what will be eaten up clean in fifteen minutes.** If there is any left over after the ducks have eaten briskly for fifteen minutes, the food boa d **should be scraped entirely clean.** When you get skillful at feeding from practice you will know just how much to feed. The idea is not to give these fattening ducks too much food, for if you do they will not keep fat. **You must keep their appetites eager.**

Make the morning feeding **light.** Make the noon feeding **light.** At night **be more liberal.** Then give them the food good and strong. This will last them through the night.

Keep them eager. They ought to go to the man who is feeding them at feeding time. Do not disturb the ducks while they are feeding. Go on about your business to the next pen. A green man standing around the food board and watching them will keep them away from their food. Scare them off and they will not go back to the food.

Always water before feeding. Give the water to these fattening ducks in water troughs built like illustration.

Give the food to them on boards the width of a shovel. Keep the water trough close to the food board. **Do not put it at the end of the yard,** for if you do the ducks will not go to it to get water, especially in hot weather. They eat at the food board greedily and they want water every now and then handy, to keep them from choking. They do not like hot water. They want cool, fresh water supplied at each feeding time, also water between feedings. That is to say, this extra watering for the fattening birds is done at 9 a.m., and 2 p.m.

You judge by the **appearance of the bird** when it is fat enough to kill. **Killing age may be at eight, nine, ten, eleven or twelve weeks of age.** Pick up the duck by the neck and feel of the body. Feel of the duck's back. If it is fat there it is fat all over. If it is thin there, put it back into a pen reserved for culls. These culls should be examined three weeks after being thrown back to determine whether they are then fat enough to kill.

MEAL TIME

Wire netting separates these pens. (It can hardly be seen in the photograph.) These ducklings are in one of the cold houses and are six weeks old. (In another week they will be ready to drive into the large fattening sheds.)

They are seen getting their food out of the boards with raised edges and their water (look closely) from V-shaped watering troughs. It is important to set the food and water troughs near each other because the birds run to the water after every other mouthful, both to get a drink and to wash their bills. In the fall, these yards, as well as the others, are ploughed and sown to green stuff (winter rye) in order to keep the soil fresh.

FATTENING

If you cannot tell by feeling of the duck whether it is old enough to kill, **put it in a bag and weigh it.** When six or eight weeks old it ought to weigh **four pounds,** or **four and one-half pounds.** At ten weeks it ought to weigh **five and one-half** or **six pounds.**

When they are really fat they ought to be killed. Keep them longer and they are going to lose some fat.

When they are eleven to twelve weeks of age they have a light moult, shedding some feathers. They lose their appetites and go back in weight a little. **They should be killed and sent to market just before this light moult starts.**

Do not feed any sour food. Do not mix at any one time more food than you need at that time, or it may sour over night.

Get the knack of feeding so as not to have to clean up the boards after fifteen minutes of feeding. A skillful man will make a mistake in this matter only about once a month.

This stuff which may be cleaned up after feeding **should be scraped off,** either into a pail or basket. It is not necessary to waste it even if it is a little sour. It may be mixed in and fed with new food, provided there is not much of it. (It is said of the duck growers in France that they allow the food to sour a little purposely for the reason that they think it fattens more quickly.)

Novices in the duck business have trouble in the brooder house with ducklings dying from sour food; watch out sharply for sour food there. If you should find any sour food there, you can get rid of it by mixing it with good food and giving it to the **older birds.**

The color of the skin of a first-class duckling when ready for market should be **white, not yellow.** This is one of the reasons our ducks bring better prices in the markets. Yellow skin is caused by too much green food, or by letting the birds stay on green grass ground too long before killing. For the last two weeks previous to killing the ducklings should be kept on **bare (dirt) ground.** The object of green food is to keep the bowels of the birds in a good, open, healthy condition. Too much eating of green and too much lying on green grass make their flesh yellow.

In our latitude we use for green stuff much **winter rye.** We buy the seed not of a seed man, because he charges seed prices, but of a grain man, at grain prices. We sow this winter rye in the latter part of September and the first part of October. Sow it in September and you have a chance to cut it twice before freezing. The ground where we plant it (the runs) is so rich with the duck manure that the winter rye grows much faster than on the average farm. It does not freeze in the winter, but stays green. If a warm day should come and the snow melt in the winter, you can cut the winter rye and feed it to the ducks. Just before we expect a snow storm we cut it and keep it frozen in a building where no sun penetrates. It will keep well. Use as necessary.

FATTENING YARDS
Road down the middle allows easy access to all pens with food and water.

CHAPTER EIGHT

KILLING, PICKING, SHIPPING

How the picker uses first a knife, then a club.—Sharp knives necessary to remove some feathers—What the picker wears, and the wages he makes. — Washing the bird. — The process of tying the head under one wing.—Plumping the duckling in tanks of ice-water.—The shipping box.—Express pointers.—Use of barrels. —Not necessary to use ice.—Distance possible to ship is almost unlimited.—Choosing a commission man.—Shipping ducklings alive.—How to scald and pick ducklings.—Getting the food passage empty before killing.—Care of the feathers, how to pack and ship them.

The killing is done by the picker, who stands, holding the duckling between his legs, takes a sharp knife, double-edge, opens the mouth of the bird and cuts the roof of the

STICKING KNIFE

The duckling is killed with this knife. The blade is six inches long and one inch wide. The whole knife, handle and all, is ten and three-quarters inches long. The back should be ground and sharpened two inches down from the point, as well as the front edge, and the point should be kept sharp.

duckling's mouth inside, making a cut of considerable depth so as to sever all the blood passages.

Then he stuns the duckling by striking it with a club. This club is generally about eighteen inches long and not too heavy. Part of a wagon stake or a hoe handle answers very well. The bird is struck a good sharp blow

HOW WE MAKE DUCKS PAY

so as to make it insensible. One blow is all that is necessary, although some pickers less skillful than others strike the duckling two or three times. The blow is aimed directly upon the top of the head with the club. The skull is not smashed.

The bird is **stuck before it is stunned** so that the blood will run better.

While the duckling is bleeding, the picker goes immediately to his chair beside the feather box, sits and begins picking. The duckling is cooling now and when the temperature of the blood gets below ninety-eight degrees **it congeals and stops running** out of the duckling. The picker's feather box is about level with his knees so that he can drop the feathers easily off the duckling into the box as he picks them. He holds the head of the duckling between one knee and the box to prevent its fluttering and soiling the feathers with blood. There is a pail of water suspended from a wire directly over the feather box and the picker frequently wets his hand in this pail. This water causes the feathers to stick to his hand, which also gives him a grip or purchase on them so that he can pluck them out without much effort. The wing and tail and other coarse feathers are thrown out, as they are too hard and rough. The picker works by making a sharp jerk in the opposite direction from which the feathers lie, the skin meanwhile being drawn by the other hand so that it is tight. If very tender, the skin at the roots of the feathers is held between the fingers, and the feathers are pulled out straight, a few at a time. The pin-feathers are wet to cause them to stick to the hand and are then caught between the thumb and the blade of the knife held in the right hand.

This knife is an important item to the picker. Each picker has from a half dozen to two dozen of these knives, and he sharpens them at noon or after work so that his working hours are not used up in the sharpening. We illustrate the style of knife used by the pickers. The blade is good steel. It is ground and honed, then stropped on an **ordinary razor strop.** Each picker has one of these leather

KILLING, PICKING, SHIPPING

razor strops hanging beside him from the feather box. The knife is kept as sharp as a razor, for part of the bird really **must be shaved** to make a clean picking job.

The pickers make from twenty to thirty dollars a week. It is all piece work. They get so much for every duckling.

KNIFE USED IN PICKING

This knife is seven and one-half inches long over all. The blade is three and three-quarters inches long, and three-quarters of an inch wide. It must be of good steel so that a razor edge can be kept on it. Each professional duck picker has from half a dozen to two dozen of these knives and it is an important part of his business to keep them sharp.

When their price is five cents for every duckling, **the feathers** pay for **the picking**. Some pickers demand eight cents a duckling. It is money well earned, and the weekly wages they make are not too much, as a degree of skill is required. It is a real trade.

Old clothes should be worn in picking. The professional picker takes off all his clothes before beginning work in the morning, puts on an old shirt and a pair of full-length overalls which are white to begin. They are first oiled with raw linseed oil and left outdoors in the sun to dry for a week, then they are given a coat of linseed oil and varnish. This makes the overalls moisture proof. They are generally worn by the picker until worn out. They are never washed. They are hard and stiff—like armor plate.

In the picking room there should be a barrel or box of lime, air-slaking. This air-slaked lime is sprinkled around the picking room on the blood on the floor, to keep the place sweet.

The picker who makes a slip and cuts the skin of the duckling, or rips it, **must sew it up.** For this purpose each picker has a needle and a spool of thread and if he makes a cut or rip he quickly sews it. When he has donet his it is almost impossible to find the place.

DUCK PICKER AT WORK

KILLING, PICKING, SHIPPING

The professional picker generally strops his knife by turning it on the strop on the edge, not on the back, as a razor is stropped. However, each man has his own way of keeping his knives sharp.

A good picker should pick from forty to fifty ducks in a day. More than fifty a day is above the ordinary. Often a skillful picker is found who will average **sixty-five a day.**

Each picker has a counter or tally device like a baseball umpire's counter, and as he finishes a bird he turns the counter.

He puts the bird, when he is done with it, into a tank filled with water. This tank is made with compartments, eight or ten of them. Each picker has his own compartment for the birds which he picks, so that his work can be checked by the foreman. The foreman, who is generally the man who ties up the birds and carries them forward to the shipping boxes, takes the birds one at a time from the picker's tank and washes them to get the blood off, and the dirt off the feet. The washing is done in an ordinary pail. It is finished in cleaner water in a second pail. The foreman then puts the duckling into another tub of water, not ice water, but ordinary faucet or spring water, to get the animal heat entirely out of the carcass. This saves ice when the ice is used later on in the process.

The foreman then **ties up the bird** and this is an interesting process, as it makes a handsome, compact duckling. There are one or two details about this work of tying which should be noted carefully. The head of the duckling is bent around and back and **put under the wing.** A string is then passed around the entire bird at the middle of the body and a common hard knot made at the wing. The feet are allowed to stick straight out. The string confines the head and wings. A forming box or press such as is sometimes used in tying fowls is not necessary for ducklings. The operator works on top of a table with his hands alone. **The string should not be fine,** like harness thread for instance. Such fine string or thread, although it may be strong, will prove a nuisance because it will cut the fingers of the operator. The string should be a good-sized white kind which can be handled easily and rapidly by wet fingers without cutting them. After the knot has been made the operator cuts the string with his picking knife. Be sure that the knot is on the side **at or below the wing.** Ducklings are often seen in the markets with the knot of their tying string directly over the centre of the breast, just the place where it ought not to be, because there it looks slovenly, and spoils the appearance of the birds.

This string sinks into the flesh and is almost out of sight and hardly shows when the flesh is soft and wet.

When the operator starts to tie up the duckling, it is lying on the board, breast up. He grasps the head in his right hand, swings it around to the right and puts it between the wing and that side of the body. The duckling after tying is put into one of the refrigerator tanks. Each tank is a convenient size, about six or eight feet long, three or four feet wide and three to four feet deep. These refrigerating tanks are made of common Yankee pine or they can be made of cypress, or any wood suitable to hold water. If you give these tanks a **good coating of paint inside** you will find that they will wash out much quicker and better and will not become slimy like an unpainted tank. Water is first put into the tank to about one-half the depth, then the ducklings are put in. As a rule the birds float in the water. Once in a while a carcass will be found which will sink. As more and more ducks are put in they press down those already in the tank. The ice of course floats on top of the water. The duck is allowed to stay in the tank of ice-water until shipping time, which is generally from twelve to twenty hours later. The object of the ice-water is to plump the flesh and condition it. The water also adds slightly to the weight of the duck.

The ducklings should be plumped in the ice-water over night and shipped the following day as the trains run.

The ducklings are shipped to market by express (not freight) either in boxes or barrels. We use boxes in ship-

KILLING, PICKING, SHIPPING

ping by express to the Boston market and **get the empties back free by express.** We use barrels in shipping to the New York market because when we ship there the shipments are handled by two express companies and we cannot get back the empty boxes at a cheap rate.

The shipping box we use is a substantial affair. A light, fragile box would not answer. **The box holds twenty-four ducklings.** Sometimes twenty-six will go in.

The inside measurements of the shipping box are twenty-four inches long, fifteen inches wide, fourteen inches deep. It is built of one-inch pine. Better and lighter wood can be found in various sections of the United States. The cover of the box is not on hinges, but is bolted on with two bolts, one at each end of the cover in the middle. These boxes are used over and over again until they are worn out. They last for four or five years. The common or merchandise rate is charged by the express companies for taking the killed ducklings in these boxes to market. For the ordinary express-train journey of six or seven hours or less, **no ice is used.** If the breeder is shipping long distances, ice should be packed in the box along with the birds. There is practically no limit to distance which killed ducklings can be shipped, as the markets are located in America, if the shipper packs correctly.

Before putting the ducklings into the shipping box **the box should be lined with brown paper.** We do not mean that this paper should be tacked in. We have a supply of brown paper such as grocers use, the sheet being long enough to go across the box. We put a sheet on the bottom of the box which covers one side, then another sheet, then a third sheet on top of the ducks after they have been put in. This brown paper prevents the ducks from coming in contact with the wood, which may be dirty. It keeps them clean and the interior of the box **has a sweet, clean look** when opened.

Shipping boxes like these should be used by anybody shipping to his market by one express company. If you have a market reached by two express companies from

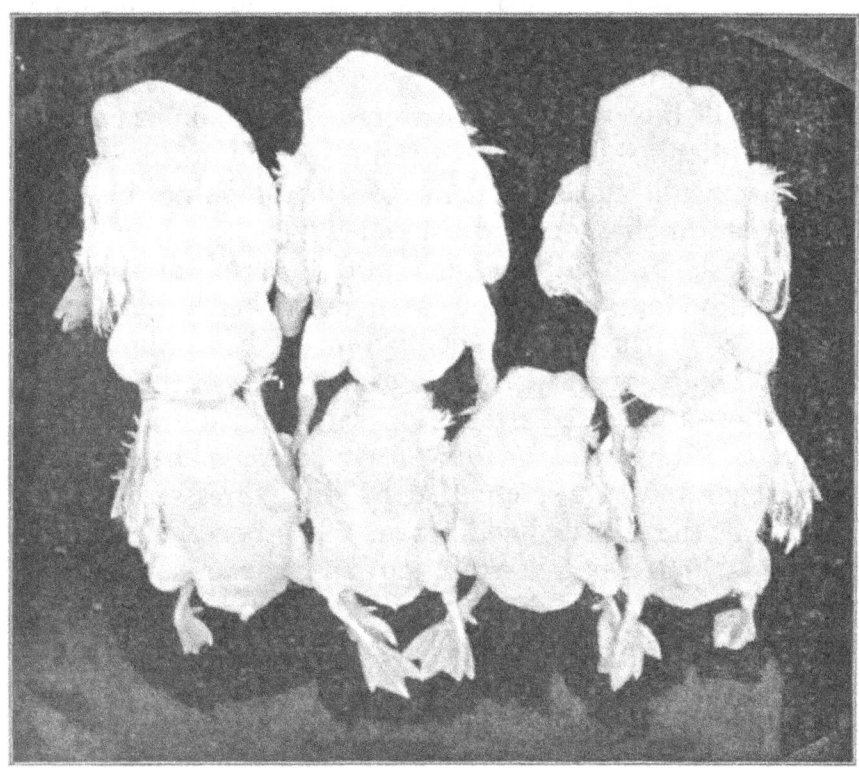

READY FOR SHIPPING

Ducklings killed, picked and ready for packing and shipping. The string which confines the head under the wing has not yet been passed around over the breast and tied. Observe the large frame, full breast and general plumpness of these birds. When the photographer took the above picture, the ducks were lying on a horizontal table and the camera was six feet distant, higher than the table. The result is, that the three ducks on the inside (or bottom) row appear larger than the four behind (being nearer the camera). They all, however, were about the same size and weight, with the exception of the two at the left of the lower row, which were exceptionally big.

These ducklings sell readily on their good looks and their delicious taste. Their plump, white flesh and trim appearance make them marked objects in any poultry or general market display.

The ducklings in the picture were eleven weeks old when killed and weighed sixteen pounds to the pair.

KILLING, PICKING, SHIPPING

your place, you will not get the empty box back free as a rule, although you might in some cases. In that case **barrels can be used,** as we use them in shipping to New York City. We use sugar and flour barrels. They cost us eighteen cents empty. A sugar barrel will hold from forty-five to fifty ducklings. We do not head the barrel, but lay the paper in, then the ducklings, and on top of the barrel we stretch a piece of burlap, tacking it around the top of the barrel. A flour barrel holds from thirty-two to thirty-five ducklings.

Sometimes an expressman, if he is green at handling these duck barrels, will turn one over and stand it on its head instead of on its bottom. This jams the top layer of ducklings but does not spoil them. In the summer we put ice in the barrel. The ice melts, as it should, because poultry keep better with cold water sprinkling over and around them than they do with only ice on them. The water collects in the barrel and if the expressman turns the barrel over it will run out and annoy him. The best way is to use boxes and not barrels if you are located so you can. We can use the boxes, shipping to New York, but it would cost us more to get the empty box from two express companies than a new barrel can be bought for.

The New York market is a very strong one. People who do business in New York and live out in the country if they want to raise ducks need not fear for the market. That city will take an unlimited quantity of anything in the poultry line. **The same is true of any other city in America or Canada.** Wherever people are gathered together there is a lot of eating going on, and anything in the poultry line is absorbed naturally as a sponge takes up water.

We never use ice in shipping ducklings to Boston. The dealers give us fair weight. We have never had any trouble with any marketman on ducklings. We have a set of scales in the shipping room and we get **the net weight** of every box as it is made up. **We allow for shrinkage** in the dressed ducklings and are able to hit it exactly right after experience. First we let the carcass drain for five minutes before putting it into the box. We squeeze it with our hands to get as much water out as possible. **For every one hundred pounds net weight of ducklings there will be a shrinkage in going to market of three or four per cent.** That is to say, when a marketman weighs them and pays according to his weight, he will return to you a weight of ninety-six or ninety-seven pounds to every one hundred pounds which you weighed.

Prepay the express charges when shipping ducklings to market. This ends the matter. See them weighed yourself at your depot, pay your own agent and he will give you a receipt. He is a friend of yours and he will weigh them correctly. If you let the shipment go forward to the city, express charges collect, you never know exactly whether the charges are figured properly at the other end. The expressman who does the weighing at the city end may be a new boy, just entered the employ of the company. The delivery-sheet writer may make an error. By the time the box gets to the marketman there may be an excessive charge. It is not for the interest of the marketman to question the charge, because it does not come out of him, but out of you. He will sign the driver's sheet as quickly as possible, pay the charges and that ends the matter except when he bills it up to you; and you have to stand for them, unless you wish to go through the red tape and delay of getting a rebate. For these reasons we say to you emphatically, **always prepay your shipments.** We would not have the reader believe that express employees are careless and inefficient. We speak of these matters as a warning, a possibility. Prepay the shipment and your mind is at peace.

The question is often asked, how long a distance can killed ducklings be successfully shipped to market? Even breeders who live in the West sometimes want to ship to New York. It is hard to answer this question in positive terms; it depends on the season, on the man who is doing the shipping, on the express company which is handling the

KILLING, PICKING, SHIPPING

shipment and on the promptness with which the shipment is picked up at destination. As a rule, we would say that a distance of four or five hundred miles, such as from Buffalo to New York City, is all right. However, as this country is laid out with markets sprinkled everywhere, east, west and south, **there is no excuse for anybody having any doubts as to being able to ship poultry.** Get to work and turn out the ducks and such matters will take care of themselves. In Chicago, while visiting there, we saw a truckload of ducks for the provision market of one of the big department stores in that city. They had come from Pennsylvania and probably had been bought in New York City. Meats and poultry are sold cheaper in Chicago than in the eastern cities, and if a firm of this magnitude in Chicago can get its ducklings from such a long distance, as we are informed it does steadily, any breeder can ship ducks profitably any corresponding distance to market.

In choosing a commission man or marketman, it is a good idea first to make him your friend. Tell him what you are going to produce. Write him or see him. Talk to him in a friendly way. Do not look upon him as an enemy. **Do not change around from one dealer to another.** When you have found a good man and got him acquainted with you and your ducklings, stick to him. His customers will praise your ducklings, they will tell him they are fine. He will write to you and say he is pleased with them. His trade in them will grow, and he wants it to grow because he will make more money. Let him push your goods. Stick to him and he will stick to you if he is any kind of a man. If you change around from one dealer to another they will not take the interest in your ducklings if they know that some other man will get the next lot of them.

The whole duckling is not picked. The wing is picked up to the first joint. The neck is picked half way up to the head. **The duckling is not opened or drawn.** You must not take out the insides of a duckling before shipping. The birds will not keep nearly so well. They will begin to

HOW WE MAKE DUCKS PAY

mould on the inside with the slightest delay. The marketmen want them undrawn and that is the way you always should ship them. The birds are drawn by the marketman when he sells them to the customer, or the customer cleans them at his or her home, hotel or restaurant, or wherever the cooking is done.

The killing and picking of ducklings can be avoided entirely, by shipping them alive. Many small breeders never ship any ducklings killed, but always alive. There are poultry gatherers everywhere who go about in wagons picking up live poultry. You will find their advertisements in the papers, asking you to write or telephone them, then they will call and get what you have. These gatherers take their goods off to the marketmen to be killed and picked. In some cases, they do the killing and picking themselves.

Some marketmen will take the ducklings alive. Write and find out whether the marketman to whom you propose to ship will take the ducklings alive.

The Hebrew and the Chinese trade in the large cities consume large quantities of ducklings. They wish them alive as a rule. Fowls have to be killed in a certain way to conform to the Hebrew religion. An excellent trade in live ducklings can be worked up with Hebrews and the Chinese.

Ducklings sold killed or alive straight to the consumer bring the money which the commission man or dealer gets. **A trade of this kind is worth working up,** for the greater profits in it.

Breeders are shipping ducklings to New York, Philadelphia and other markets outside of New England, which have been **scalded and picked.** When a duck is scalded before picking, the feathers come off much easier. Not over five cents is paid for picking a scalded picked duck, because it is much easier work. The tail feathers are left on.

A dry-picked duck in any of the eastern markets is always called a Boston duckling.

KILLING, PICKING, SHIPPING

Many city markets see nothing but scalded ducks. The process of scalding is, first to immerse the duckling in a wash-boiler of water that has just come to a boil. The water must not be hotter than this. The duckling is held by the head and feet for half a minute in the water, then the feathers are immediately picked off. It is quite common for women who have a few ducks and ship only a few at a time to scald them before picking.

To show you the present condition of the New York market with regard to scalded ducks and dry-picked ducklings, it is true that **for every one hundred ducks marketed in New York City ninety-five have been picked after scalding.** It is more trouble to dry-pick them in the manner we have told in this chapter. **Scald or dry-pick as your trade demands.**

Some picking is done before killing, generally in midsummer when the feathers come out easier. This is a cruel practice which hurts the birds, and we do not think it ought to be done.

We once had a man in our employ who claimed to be able to kill a duckling so that the feathers would come off easier. His theory was that he had found a certain spot in the brain of the duckling which when he ran his killing knife into it affected the nerves of the whole body of the duckling so that the feathers were, as it were, released by the duckling. This seems silly to read but it is a fact that this picker got his feathers off more quickly than his fellows. He always ran the knife into the brain of the duckling from a peculiar angle. We do not vouch for the value of this information, but simply print it as a bit of gossip of the pickers.

Ducklings about to be killed should have their last food at night **so that their food passage is empty when killed** the next day. They can be given plenty of water to drink before killing, but if the food passage is filled with grain when the bird is killed, this grain will ferment, sour, turn green and spoil the flesh. If by mischance a duckling is killed which has eaten and filled the food passage with

DUCKS IN THE FIRST PEN MOVED IN THIS PICTURE, CAUSING THE WHITE STREAKS

KILLING, PICKING, SHIPPING

grain, the neck should be squeezed and the grain washed out through the mouth before shipment.

The ducklings do not lose weight between their last feeding and killing time provided they have all the water they want to drink.

A large plant will kill and ship on an average two hundred ducklings a day, when busiest four hundred a day. An average of three hundred fifty to three hundred seventy-five a day will keep nine pickers at work.

Every twelve ducks will give up **a pound of feathers worth on an average forty-five cents.** This price may vary in different parts of the country. We have been getting **fifty cents a pound** this year — more than ever before. There are feather buyers everywhere. Their advertisements may be seen in many journals. Write us when you are ready to ship feathers and we will tell you a good place to ship them.

The feathers are taken from the picking room several times a day and put in the feather loft. We use the loft of a grain barn. We throw them on the floor of the loft and stir them up with a pitchfork once a day for three or four days. By that time they are dried at the roots and can be pushed together into a pile to make room for new feathers from the picking room. Turn the pile over with a pitchfork once a week. The idea is to get the feathers loose. **Do not let them pack up and get heated, and ferment.**

To ship the feathers to market, **use bags** made of white cotton cloth. Formerly we made the bags at our own expense, then we found out that the feather men, if we asked them, would send us the bags, so we got rid of this expense. Write to your feather man, tell him what you have, and he will send you the bags. A feather bag is generally six feet long and two and one-half feet wide. The feathers are packed into it tightly by hand. Sew up the top of the bag with string. **The weight of a bag packed properly should be close to sixty pounds.** It takes twenty to thirty minutes to stuff a bag. A bag may be

HOW WE MAKE DUCKS PAY

filled easiest when it is suspended beneath a hole which has been cut in the floor of the loft. The feathers are then pushed, packed or shovelled in more easily than if picked up by handfuls.

The feathers have a little odor when shipped. The feather man takes off this odor by using first a steam renovator which dries the feathers and kills all the animal germs in them with superheated steam which is very dry. The feathers are then put through a blowing machine which separates the down. The feather men get more money for this down than they do for the feathers.

These feathers are used to make bed and sofa pillows and all kinds of pillows. They are also used to make beds, especially for foreigners from Europe, where feather-beds are much more in use than in this country. There is quite a trade in these feather-beds, old-fashioned as they are. The demand for feathers for pillows never lets up.

PEN OF DUCKLINGS ON DIRT RUN A WEEK PREVIOUS TO KILLING

CHAPTER NINE

MARKETS

To learn the true market prices of ducklings, ask the dealers what they want for selling to you, not what they will give when buying of you.—Letters from remote sections of the United States and Canada showing the wonderful opening for our ducklings.—Encouraging reports from everywhere, with the names and addresses of the writers. —Interesting information from the middlemen.—The best way to sell.—Facts given by dealers.—Demand for our big tame ducklings is admitted to be great, with assurance of high prices, even in those States where grain is very cheap.

The best way to find out how the market stands is to ask the commission man or dealer for what he is **selling** ducklings. Don't tell him you have, or may have, some to sell. **Ask to buy some.** Then you will learn the real facts about the market. With that information in hand, see the commission man and **tell him you will sell ducklings at that price,** less his five per cent commission.

There is considerable foolishness in the commission business in this way. Some of the commission men claim to be handling farm produce on purely a commission basis, returning to the shipper the full price received, less ten per cent commission. This is not always true. They are not satisfied with ten per cent profit. **They buy as low as they can** and take ten per cent off of that, then they sell for what they can get and this selling price represents a profit of from thirty to one hundred per cent.

In various sections of the country where we have sold breeding stock, we have written to commission men and poultry dealers (whose names we could learn in no other way than by looking in city directories), in order to find out what they would pay for ducks. We have done this at remote points, as we had plenty of knowledge as to the immensity of the big city markets.

We have always found by such inquiries that ducks are sellers everywhere, and we know that ducklings bred from our stock would go like wild-fire anywhere.

We recall one customer in the vicinity of Atlanta, Georgia. In writing to three wholesale dealers there, **in October, when prices for ducklings are the lowest of the year,** we received the following replies:

M. D. Thompson, 42 Madison avenue, Atlanta, Georgia. —"We are selling now old ducks that dress fat for **sixteen to eighteen cents a pound,** and this demand will continue throughout the poultry season. **We could place stock of your quality to the extent of eight hundred to one thousand pounds a week.**"

McDonald & Co., 43 South Broad street, Atlanta, Ga.—"Ducks such as you breed usually sell here at **sixteen to eighteen cents** a pound. Our market prefers scalded stock. However, dry-picked stock keeps longer and better and can be sold all right. Poultry is sold here with heads and feet on, undrawn. Crops must either be empty or drawn. We will be glad to serve you."

Phillips Bros., 4 North Broad street, Atlanta, Ga.—"Yes, we can use your ducklings at ten or twelve weeks old weighing five to six pounds, and sell them for **sixteen cents to twenty cents** a pound. If this class of ducks or ducklings can be produced, it will be a **darling** for the purchaser. We would be glad indeed to see them and you can count on us as being a **heavy purchaser.** Let us see some of those ducks. Only prove your assertion to us and you need not go further South. It is good enough for us. The above may seem to you that we are doubtful of your production, but such can be done. Being from Missouri, we want to see it."

If the above dealers are ready to pay such good wholesale prices for good ducklings, the outlook in the South for a high retail price is indeed excellent.

MARKETS

In the case of another customer (in Kentucky) who negotiated with us for a large purchase of breeding stock, we looked up the markets there and received the following encouraging letters:

Carson-Proctor Co., 136 West Jefferson street, Louisville, Ky.—"We handle a great many common puddle ducks, which are about all that is used in our market. In case your customer should purchase a few of your ducks for our market, we would be very glad to take the matter up with him and handle his ducklings at market prices."

H. M. DeBow & Co., 206 East Jefferson street, Louisville, Ky.—"I know nothing of the quality of your ducks, as I have never handled any of them. We have a good demand for our home ducks and **can sell all we can get** at prices ranging from eleven to thirteen cents per pound. Our ducks weigh from three to five pounds. Both white and colored are in good demand, though the white is more salable than the colored. It might be your ducks would command **good prices** in this market and a **good trade** could be worked up for them here. I do strictly a commission business; handle all produce on five per cent commission, make no advances at all."

Harshfield, Hile & Co., 251 East Jefferson street, Louisville, Ky.—"We handle poultry exclusively on a straight commission basis and could sell to advantage at good prices, **large quantities** of ducklings such as you describe. **Live ducks** are sold exclusively on this market. **The fowl as described by you would certainly prove a profitable investment, and we trust that you will succeed in introducing them in this locality.**"

In the winter time, in Texas, the hunters bring in to the markets wild ducks and sell them to the dealers for forty cents a pair. These ducks, small, skinny and rank-tasting as they are, sell readily to families. Tame ducklings bred from our stock would be a revelation to Texans accustomed to eating the fishy, small wild ducks. Many of the Texas dealers wish the tame ducklings shipped to them alive. We append the following letters which we have received from Texas middlemen. Remember the prices they quote are what they wish to buy them for, not what the consumer is asked to pay:

Pabst Bros., Galveston, Texas.—" You may refer your customer to us for handling ducks **picked or alive,** but much prefer them **alive,** as it gives us more opportunities for sale, hence can get better results."

F. J. McFarland, Central Depot, Houston, Texas.—" I have looked into the duck market a little and think that I can place your stock for you at a reasonable price, as the market at this point is **very short of such stock.** I have one or two places that I can place from five to ten dozen per week."

F. E. Jones & Co., 1010 Commerce avenue, Houston, Texas.—" Ducks are generally in good demand in this market. Should we be favored with any of your consignments we will sell them to the very best advantage, charging five per cent commission for handling them."

Frederick Produce Company, 908 Commerce avenue, Houston, Texas.—" Ducks are **very high** in this section during the months of November, December and January. We buy only live stock. Let us know if you have anything to offer at this time and oblige."

W. F. Puls, 704 Preston avenue, Houston, Texas.—" I believe I could work up a large trade in them eventually and get better prices and pay more. I have the best place in the city for dressed poultry and handle more of it than any firm here. The only trouble I have is getting choice poultry. There is no doubt in my mind but that ducks **can be worked up to a large business** providing the producer does not want too large a price for his ducks and if the cost of raising them, that is, the cost of the food, is not too high."

It will surprise Western people to learn that ducklings are shipped to the New York market from as far west as Iowa. We are in receipt of a letter from H. S. Webber, Sioux City, Iowa, stating that the duck breeders there are shipping to New York steadily. It is Mr. Webber's

MARKETS

opinion that he could ship a great many more ducklings to the New York market provided they were the equal of the ducklings now being marketed in New York. Duck food is cheap in Iowa and the whole middle West, so much cheaper than in the East that the express on the killed ducklings from the Central States to New York would not amount to much in comparison.

Six or seven years ago, the farmers in Illinois and other neighboring States received only six and seven cents a pound for their ducks, alive. Now they receive twelve and thirteen cents a pound, live weight. Consumers in the West have found out the fine quality and flavor of properly raised ducks. That is one reason why prices have increased as they have in the East the past few years.

Ducklings are handled by the Iowa dealers both alive and dressed.

In Minnesota and Wisconsin the duck markets are very good. We have received the following letters from that section:

Willard Hotel, St. Paul, Minn.—"There is a good market here for your white ducklings weighing five to six pounds when ten or eleven weeks old. We would like to buy them."

L. Eisenmenger Meat Co., St. Paul, Minn.—"During the summer months we have been buying our ducks, as well as other poultry, **alive,** and we dress them ourselves. Of course, now as the weather gets colder **we prefer to have them dressed,** and are urging our shippers to that effect. At this time we are paying ten to eleven cents per pound for them dressed, delivered. We will use about five hundred to one thousand pounds a week from now until the holidays. Should you have some one in this district who is handling your ducks, or who intends to have them, we should be glad to have a sample shipment. Then we will be able to know of its good qualities, and can make prices accordingly."

G. L. Bradley Company, Minneapolis, Minn.—"We handle a great deal of poultry. Nice white ducks are selling for nine and one-half to ten cents alive. We usually have a good market after Christmas. Ducks always sell better that season of the year. **We can sell any amount of ducks every week. Of course extra nice fine ducks would sell at a better price.** The kind you speak of ought to bring twelve and one-half cents and later they might sell at fourteen and fifteen cents dressed. We do not know what we can do with this class, but it is our impression that **they will sell well.** We should be glad to handle any shipment you may favor us with to the best advantage possible and give you prompt returns."

Chapman & Eissler, 275 Broadway, Milwaukee, Wis.— " At present ducks are selling at eleven cents a pound alive. The trade has not taken much to dressed ducks as yet. Later we expect to be able to sell them dressed at a good market profit. Will pay fourteen to fifteen cents per pound and can sell lots of them."

From a personal knowledge of the San Francisco market (to cite one instance of many), we are aware that most ducks for sale there weigh only four pounds. There are thousands of wild ducks in California but they weigh only six or seven pounds to the pair (from three to three and one-half pounds apiece) and they taste nothing at all like our grain-fattened, tame ducklings. They are rank, tasting oily, fishy and gamy, and cannot be sold in competition with the tame, grain-fed duckling. We have the following letter from a San Francisco firm:

California Poultry Co., Market and Eleventh streets, San Francisco, Cal.—" We are glad to see that you have a certain party to go into the duck business, and we will certainly use some of his ducks as soon as he gets them ready for the market. Prices are running from **$9** to **$10.50** a dozen. We use from five to twelve dozen every week and **like to get them shipped in alive.**"

The markets in San Francisco and other cities on **the** Pacific coast are great ones for ducks, and big money **is** going to be made there by duck breeders. The people on the West coast spend their money freely and have the best

MARKETS

of everything for their tables. These fine ducklings would make a sensation there in the eating line on account of their novelty and worth. They are not known there as they are in New York and Boston and will "go" in great shape.

We had a customer in the vicinity of Los Angeles, and he found out that the market there is a steady and paying one, as the following letters show:

H. G. Spaulding, 308 Temple street, Los Angeles, Cal.—"We are using about two dozen ducks per week, present price twelve cents per pound. Of course, we use larger quantities over the holidays."

J. W. Harris, 122 East Seventh street, Los Angeles, Cal.—"I could use two hundred ducks per week alive from December 1 to May 1, at sixteen cents per pound. From May 1 to December 1 we buy at twelve to fourteen cents per pound."

Marin's Creamery & Poultry Yards, Figueroa street, Los Angeles, Cal.—"We can use all the ducks shipped us. They must be shipped alive. The price, of course, will vary from week to week."

The demand is insistent in Colorado. The prices quoted by the following wholesalers are about half what the consumer has to pay there:

Dickey-Farney Co., Denver, Col.—"There is a greater demand on this market for ducks of the description which you give than your prospective customer **could raise in two or three years.** At the present time we are paying twelve and a half to thirteen and a half cents a pound for ducks, and **could very easily use a couple of thousand pounds a week,** and the trade **could be very easily increased by the superior quality.**"

Pinkett Bros. Commission Co., 1625 Market street, Denver, Col.—"We are the largest dealers in poultry in this city. We offer ten cents a pound for live ducks, delivered here. Dressed ducks are selling here today at thirteen cents per pound. Our commission on dressed poultry is eight per cent."

HOW WE MAKE DUCKS PAY

There is plenty of money in Montana for a nice table delicacy. The following letters are from Butte City:

N. Davis, 1126 Arizona street, Butte City, Mont.—" We have no regular prices for ducks. We get them as cheap as we can and we pay market prices. **We have calls nearly every day for ducks and drakes for breeding purposes.**"

Jonas Eisler, Butte, Mont.—" The average young duck that is in demand here is about a three to three and one-half pound duck."

Fine poultry goes just as well in any part of Canada as in the United States. **There is no duty on our live breeding stock shipped to Canada.** The Hon. Sydney Fisher, Minister of Agriculture for the Dominion of Canada, has expressed his regrets that Canadian farmers have paid so small amount of attention to duck raising. Before selling to Canada customers in the vicinity of Toronto, and in Manitoba, we made inquiries with the following results:

T. H. Smith & Co., 70 Colborne street, Toronto, Ont., Can.—" With reference to your ducklings, we presume the quality would be a little beyond the average, and would bring a larger price. We will be pleased to have your shipments at any time, and will give them our best attention."

F. Simpson & Sons, 734-738 Yonge street, Toronto, Ont., Can.—" We cannot give you much information concerning ducks as we have never gone into the business very extensively. The **wholesale price** just now is **fourteen cents** per pound."

Gallagher-Holman LaFrance Co., Winnipeg, Manitoba, Can.—" Concerning ducklings would say we think the prospect **very bright** for a proposition such as you mention. Just at present the wholesale price of ducklings is twelve and one-half cents; this we consider is about the **lowest price** which they can be obtained at during the year. With reference to the supply might state **the supply is not sufficient for the demand,** excepting at this season of the year (October). **We think there would be no difficulty**

MARKETS

in finding a market for all the ducklings your friend could raise."

The beef wholesalers handle an enormous quantity of poultry. We have the following letter from a leading one:

Morris & Co., Poultry and Egg Department, Chicago, Ill.—"We are at all times in the market for fancy ducks. There is no limit to the quantity we could use."

What is true of the above firm with regard to the demand for ducks is also true of Armour & Co. and the others of the great beef wholesalers.

The best ducklings now in the Chicago and St. Louis and surrounding markets are shipped there from New York. Anybody getting in now with our ducks in the Central States and shipping to Chicago and the other cities there will have abundant cause for congratulation.

George S. Sloan & Son, 124 South Water street, Chicago, Ill.—"There is a good demand here in ducklings you mention, we are now getting them from New York, selling from five to ten barrels a week. Prices ranging from **fifteen to eighteen cents** per pound, f. o. b. Chicago. Would want ducks dressed, scalded or dry picked, undrawn, heads and feet on."

Wellington Hotel, corner Wabash avenue and Jackson Boulevard, Chicago, Ill.—"We cannot say definitely as to quantity of your ducklings we may use, the price we will pay, etc. We are always looking for A1 poultry and should any one near us breed your ducks we could no doubt use quite a number every week."

Even the Utah market is looking forward to better duck times. Note the following:

W. M. Rash Co., Salt Lake City, Utah.—"At the present time we are paying ten cents per pound, live weight, for ducks delivered at our place of business. The demand is not heavy, being principally confined to the Chinese restaurants. **If the stock is fancy and steady supply sure, we might work up considerable business in other quarters.**"

J. M. Christensen & Co., Salt Lake City, Utah.—"We

HOW WE MAKE DUCKS PAY

are not handling ducks now but if we could get a good and steady supply, we would like to try to work up a market."

Other letters follow, as we have received them from other sections of the country:

John Blessed & Son, 347 Woodward avenue, Detroit, Mich.—"Ducklings would be worth to us from fourteen to sixteen cents a pound, scalded and picked."

Faltis Market Co., Detroit, Mich.—"We bought several thousand dollars worth of ducklings from New York this and last season and of course will be glad to use the same grade if produced nearer home. We use a large quantity beginning early in the season and do not get any good ducklings in this vicinity until about this time of the year (October). Regarding price, we would no doubt be willing to pay **on the New York basis**. As we said before, we purchase a great deal on that market."

Edson Brothers, 110 Dock street, Philadelphia, Penn. —"If the ducklings which are produced from your stock are as you say a wonderfully large, juicy bird and are much better than the average Long Island duck in the market, **we think we could work up a pretty good trade on them**, as the prices which are prevailing now on White Pekin dressed ducks are **very attractive** to the trade, and we think more of these ducks have been consumed this past season than ever before. Our market on Long Island ducks at the present time is from sixteen to eighteen cents as to size and quality. The larger ducks are most in demand during the summer months when the hotels are using them."

P. S. Dewald, 344 North Front street, Philadelphia, Penn.—"We have been using a great many ducks this fall, more than usual owing to their low price. Just at present we are getting the last of a flock of fifteen hundred head for which we paid thirteen cents a pound and at the same price could probably handle some more. It is reasonable to suppose that the price is likely to improve."

Hotel Havlin, Cincinnati, O.—"We can always use ducklings, such as you breed, when they are fresh killed

MARKETS

and dry-picked. At present we are using about ten pairs per week. This is no criterion, as ours is a new house."

Charles H. Keith & Sons, 554 West Sixth street, Cincinnati, O.—"The duck stock which you breed is rather new to this market. We had several inquiries last spring, but it was so late in the season that we gave them no attention. As we have not sold any white Pekin ducklings, it would be hard for us to say how much we could pay for them. Should we be able to use any, **we would want them alive.** We are getting a fine line of white ducklings that dress up to four or four and a half pounds, that sell well. Later we will get celery-fed ducks from Wisconsin, that weigh about six pounds each when dressed. **Tell your prospective patron to start in and raise his ducklings.** Just before they are ready for market let him compute the price at which he can sell them, and communicate with us."

Rowland Poultry Co., 2005 Pacific avenue, Tacoma, Wash.—"Yes we can sell your ducklings. There is a good sale during the winter season."

As a rule, most of the poultry markets in the United States and Canada do not know yet what a good duck is. **There is a splendid opening everywhere for breeders with the right birds.** Get into your nearest market, capture your share of it and get the good prices which your ducklings will bring.

What is known among epicures as a canvasback duck is a wild duck from the breeding grounds of Chesapeake Bay. They live largely on the wild celery which grows there. They weigh eight pounds to the pair. They are much prized by many diners on account of their peculiar flavor due to the wild celery and are worth about five dollars a pair in the markets.

Most of the dealers whose letters we quote in this chapter know only the common or puddle duck, weighing only three or four pounds when full grown. At three months they weigh only between three and four pounds. The eggs are small and greenish in color. The Pekin eggs are large and white. It takes the whole summer for the puddle duck to mature. Compared to our Pekin ducklings they look like a sparrow alongside of a chicken.

It is not uncommon to see in the markets small, thin, bruised, half-fattened, half-picked ducks. Many of them have bloody bills and their feet are dirty with caked manure and mud. Avoid sending to market anything in that class. Open a box of our ducklings and you see first the brown paper, a good introduction to the contents, then the contents themselves, clean of bill and feet, white and plump, something good which whets the appetite and makes one long for possession. Ducklings properly marketed give the buyer a good impression.

The red-head duck (wild) is thought to be good eating in some sections, Maine for example. A friend of ours killed a red-head down in Maine which weighed four pounds. Three trout were found in her gullet. She was roasted with all skill at command but tasted oily and fishy and was a disappointment.

All the letters from dealers which we quote in this chapter were written by them in **1906, summer and fall. At that time, everywhere, ducks are at their lowest price. The figures given in the letters may be considered the minimum prices of the year. The spring prices are much higher.** We have not investigated the markets much in other sections of the country except in a general way, being advised by customers that they were selling profitably everywhere. There is not a place where ducklings cannot be raised and sold at the same profit we make here in the East, because it is all a question of finding men, women and children who like table delicacies. **These people live around every market.** Ducklings are as salable as anything eatable they buy.

Lake, ocean and river steamers, dining cars on all railroads, hotel and restaurants everywhere, clubs, etc., all want good ducklings. Sell to them direct if you can and make the fifty to one hundred per cent profit which a middleman will make if you sell to him first.

MARKETS

Commission men, poultry dealers, hotels, etc., are pestered continually with letters from poultry experimenters and dreamers. Most of them pay no attention to letters written by curiosity seekers and throw them into the waste basket. They are always interested to talk business with anybody **who actually has poultry to sell,** and proves it by his letter. Anybody who starts off on a campaign of letter-writing or walking tour of investigation as to whether ducklings are salable, and at what prices, would better, as we have suggested before, inquire for what he can buy them. Let the dealers come to you, when you have the ducklings ready. If you are breeding the right ducklings, they will drum you for the chance to sell them. It is all under your control.

Make a start, turn out the ducklings and begin learning the markets as well as other details by actual practice. **An ounce of this practice, this actual handling of the business, is worth a ton of theorizing.** The subject takes on a near and real aspect. We have had customers write four-page letters for weeks asking full details about size of shipping boxes, locations of markets, names of marketmen who would take one hundred ducklings a day, etc., fearing that should they embark in the industry they would flood their nearest city with ducklings which would be a drug in the market — and all the while these beginners did not have even a trio of ducks; their fears existed on paper only. Anybody who can entertain doubts that ducklings and other poultry can be sold profitably when raised has not intelligence enough to succeed in poultry raising. Such people should face at the start the fact that they are unfitted for business on their own hook and should keep on working for others more resourceful and more enterprising. It is weak and pitiful, when a man presumably intelligent and at the age of discretion will write and say: " I live in a small place two hundred miles from any city and I don't think I can market ducklings if I raise them, or make any money with them; do you believe I can?" What can be said to such a man to convince him? Can anything be said briefly? Hardly. Such a man must be educated from the beginning. He has no imagination. He cannot conceive that there are people who like to have good things from day to day on their dinner tables: city people, and country people too, not only the wealthy ones, but the comfortably well-off, who are searching for nice, appetizing food to eat all the time. Never having been in a great city market, he does not realize that tons and tons of ducklings and car-loads of chickens and eggs melt away there every day like dew in the morning sun. We speak of this subject emphatically because it is an exasperating experience to receive a letter from a beginner expressing doubts as to the markets, and fears that he will flood his nearest market, once he starts. Such letter writers almost invariably state their alarm that the whole country soon will be raising ducks, and that ducks will drop in value to nothing.

Another doubt of beginners is that they cannot sell ducklings except cheaply to commission men and dealers. **Why should a duck breeder sell to a commission man or dealer, if he does not wish to?** The selling of his product is always under the control of the breeder. He can sell to whom he pleases and is not obliged to take the first offer. We have always sold to commission men and dealers and made a good profit, but we have been well aware all the time that we could have made more money selling over their heads direct. Some dealers and commission men will not pay what ducklings are worth if they find they can impose on the breeder, or keep him in ignorance of the market.

New York and Boston will take all the ducklings offered at what are now **the highest prices in America.** Those of our customers who raise ducklings in New England, and in New York, Pennsylvania and other States within shipping distance of New York and Boston are more favored than those in remote States, but **the day of the latter is coming quickly.** An idea of what the Boston market is may be obtained from the following:

MARKETS

Nathan Robbins & Co., 33 and 35 Faneuil Hall Market, Boston.—"We find that we sell as many if not more ducks than chickens and they are daily becoming more popular. The public is beginning to realize that ducks are delicious eating and when properly raised are very meaty. The average price that we pay for our supply of ducks during the year is about **seventy-five cents** a duck. They weigh between five and six pounds. **Of course during March and April, when the new ducks come in, we have to pay double this price.** First-class ducks such as are raised from your stock always bring the high prices and are easily sold. We find that the prosperous, first-class trade will always pay way up for the best ducks. We use a great many ducklings during the summer and store away a large number for winter trade."

Swan, Newton & Co., 18 Faneuil Hall Market, Boston.—" We can sell your white Pekin ducklings up to the limit of production, and good ducklings of this breed never go begging in this market. The native Pekin ducklings that we put away in cold storage for winter trade sell much better and bring higher prices than the so-called young, fresh Western ducks shipped here during the winter. We buy, however, quite a number of the western ducklings during the winter as the supply here is scarce. There is no special season for the sale of ducks. We find the demand on the increase and good the whole year. During March and April, when the green or new ducks come in, the supply is very limited and they sell in open market for **thirty-five cents per pound.** The earlier the dealers get them in the market, the better the price they get for them."

Adams, Chapman & Co., 38 North Market street, Boston.—" We handle more ducks than any other firm in Boston, and are in the market at all times **for all we can get.** We pay from **fourteen to eighteen cents a pound for them and have never found the supply equal to the demand.** Many of the duck farms are increasing the size of their output, owing to the increasing demand and the large profits in the business. We get twenty-two cents a pound for cold-storage ducks during the winter. The western ducks shipped to the Boston market are old birds weighing three, four, and five pounds. They are not to be compared with the fresh, native ducks bred around here. We can sell all the ducks that are sent in to us."

Harrison, Swan & Co., 1 New Faneuil Hall Market, Boston (Mr. Richardson). — " We buy large quantities of Pekin ducklings at ten weeks which weigh between five and six pounds apiece. We buy direct from the duck farms and also from commission merchants in this city. During January these ducks bring **thirty-five cents a pound,** and during July and August, which are the lightest months of the year, **sixteen cents and seventeen cents per pound.** They seldom sell here for lower than this. There is an active demand for good, meaty ducks and **we can sell all we can get of them.** One large grower with whom I am acquainted averages the entire year about **one dollar per duck."**

J. P. Lawrence & Co., 30 Faneuil Hall Market, Boston (Mr. Woodman). — " There is an excellent demand for white Pekin ducklings and **we easily sell all we can get** at a good price to hotels, restaurants, and private consumers. We make from three to five cents per pound on every duck we sell, and the raiser much more. During the summer we put many ducks away in cold storage for sale during the winter, when the supply is scarce. We buy most of our ducks direct and pay cash for them."

H. L. Lawrence & Co., 46 and 48 Faneuil Hall Market, Boston.—" We pay for white Pekin ducks on an average of fourteen cents per pound, and sell them for twenty-three and twenty-five cents per pound. During January and February they sell for twenty-five to thirty-five cents per pound and we have to pay for them twenty cents to thirty cents per pound. **We have never known the supply to equal the demand, which is good at all times.** Fine Pekin ducklings will always sell and people will pay the top price for them."

MARKETS

C. R. Corwin & Co., 2 Faneuil Hall Market, Boston.— "We have been paying the duck raisers this year thirteen cents and fourteen cents per pound, while last year we were paying them seventeen cents and eighteen cents a pound. **The demand for Pekin ducklings has always been good and we can sell all we can get hold of all the year round.** We are commission merchants and pay for the ducks at the prevailing market price, deducting only five per cent of the proceeds as our profit for selling them."

W. E. Legg & Co., 6 South Market street, Boston.— "We buy our Pekin ducklings alive, and can dispose of a fair amount at good prices. There has been a very fair demand this year for these birds. Duck raisers can make good money selling their stock at fifteen cents per pound and the price paid here is seldom lower than that. Last year we paid seventeen cents and eighteen cents at wholesale for our supply. **Send your live ducks in.**"

Mr. R. Norton Crane, of the Mississippi Agricultural College, writes: "How few farmers throughout our Southern States grow ducks for market in any quantity whatever. Yet in no other branch of the whole poultry industry do we find such large profits coming so regularly and so surely. There is at all times a steady demand for well-fattened young ducks on every market of the United States. **I tried Mobile last fall with some very inferior ones several months old, and weighing only two and one-half pounds, but I received twenty-four cents a pound for them.** This is paid for poor stuff. What could we get if we had first-class young ones?"

We furnish the breeding stock to breed first-class ducklings, and we would like to get into communication in the South with duck farmers who mean business and will profit by the above suggestion from Mr. Crane.

To settle the matter of duty or no duty on our breeding stock shipped to Canada, we wrote the collector of customs at Montreal and received the following reply from him: **"In reply to your letter of inquiry, I beg to inform** you that live poultry, thoroughbred, imported into Canada for the improvement of stock, is admitted free of duty. There must be produced with the entry at customs a certificate of purity of breed from the breeder of the poultry. (Signed) R. S. White, Collector."

When we ship our breeding stock to Canadian customers, we send a certificate of purity of breed, as called for, and the customer pays no duty. More letters follow:

Bruno Franz Sons, 447 Franklin avenue, St. Louis, Mo.—"We have never handled the Boston ducklings on this market. We are selling heavy young broiling ducks and have been using the common spring ducks, and if we can get first-class ducklings we can get a fancy price for them as we have the cream of the trade. We would want ducklings scald picked, and could handle several thousand pounds a month. We sell more city dressed ducks and geese than any other two houses in this city and can get fancy prices for fancy stock."

R. F. Brandt, Sheriff Street Market, Cleveland, O.—"We are in the market at any time for choice ducks or other poultry. We can use almost any amount shipped us; prefer scalded ones. Can use them alive if no convenience to dress. We consider the market good for choice ducks."

B. S. Schermerhorn & Co., 316 South Los Angeles street, Los Angeles, Cal.—"Most of the ducks come in here as grown birds and are sold alive. Selling values vary at from about ten to fifteen cents a pound, live weight. The quality or breed of the bird makes but little difference, so long as the bird is living. We handle very few ducks, but believe we could work up a nice business on fancy, young live ducklings. We certainly would like to give it a try. We believe we could obtain a premium on present prevailing prices. We believe a man could make a nice living here, running a small duck farm, stocked with fancy stock and feeding them on grains, etc., rather than on mud, water and small fish."

CHAPTER TEN

QUESTION BOX

Queries and replies.—Holding the head when killing.—Restless at night.—How not to scald.—Do not grow celery to feed to ducklings.—Dangers of going too fast, and the importance of getting experience.—Demand in remote sections.—Not necessary to kill, they can be sold alive.—Farms that raise part of the ration.—Why ten to twelve weeks is the proper time for killing.—A suggestion as to worms.—Regrettable experience in breeding to satisfy a false demand for small ducklings.—Points as to feeding and other details brought out by many letters.

Q.—**You say that when you are picking the duckling after sticking you hold the head between your knees. I don't see how you can get the feathers off if you do that. I have tried it and the bird flops around and I have difficulty in getting the feathers off.** *A.*—Please read that chapter on picking again. We do not write there what you say we do. We tell you to hold the head of the bird between one knee and the feather box. This leaves the body in your lap, where you can turn it around to suit yourself as you pick it. You hold the head tightly against the box and this prevents the bird from flopping around and soiling its feathers and body with blood. Sit in a chair while you pick.

Q.—**My ducklings have a yellow tinge to their flesh when I ship them. What is there in the food which causes this?** *A.*—This is caused by allowing the ducklings to lie on green grass before you kill them. You must take them off the grass range one week or so before you kill them, and put them onto dirt. The yellow color seems to go through the feathers to the skin.

HOW WE MAKE DUCKS PAY

Q.—**My flock appear restless at times during the night and do not always go to bed and sleep. This worries me a good deal. Do you think they are sick?** *A.*—They do not act like most animals when night comes. It is perfectly natural for them to be restless at times and move about more or less. During the day they like to sit motionless at times with heads under the wing.

Q.—**My marketman says to scald the ducklings before taking off the feathers. Now, why can't I put the whole bird right down all over into my wash-boiler? You say to hold the bird by head and feet and scald only the body.** *A.*—Do not scald the bill and feet because if you do you will discolor them. You also will take the feathers off the head. This you must not do. The feathers are left on the head.

Q.—**I am going to grow celery and feed it to my ducklings and get a higher price for the birds. Why don't you do it?** *A.*—That is a foolish idea. It is true that wild ducks which feed on wild celery bring better prices because of the improved flavor, but if you raise celery in a garden, you can get ten times more money for it as it grows than in the form of duck-meat flavor. It is expensive flavoring. It is also a failure as flavoring because a celery-fed domestic duck does not taste anything like the wild canvasback duck fed on wild celery, and cannot be substituted for the canvasbacks. A big, grain-fed duckling is better eating than a canvasback and will bring as high a price when people get better acquainted with them. Canvasbacks are sold at high prices because they are comparatively scarce and because fancy eaters have been in the habit of paying high prices for years for them. The demand has been cultivated by restaurants and hotels along with the terrapin and champagne demand.

Q.—**It seems to me strange, if there is so much money in ducks as you say, that instead of marketing forty-five or fifty thousand a year as you have done, you have not marketed one hundred or even two hundred thousand a year, and got rich much quicker. If your figures are** correct, and the work, as it seems to me, is only a question of hired help, why have you not pushed the business harder? I am of good business ability and I see no reason why I cannot accomplish in five years what might otherwise take twenty. Please advise me. *A.*—Go slowly. If you have been "figuring," tear up the paper and listen to reason. Is not experience worth acquiring? If ducks could be turned out like a brick-making machine, it would be necessary only to speed up the machinery and work day and night. But they are living things and have to be nourished and cared for. A man is busy and has quite a good job on hand when he is shipping one thousand ducklings a week to market, making a net profit of five hundred a week on them. If one ships one hundred thousand ducklings a year to market, he must erect more buildings, employ more help and be busier. To make a success, a man must keep things under his control. Don't bite off more than you can chew. What is the use of trying to do three years' work and make three years' profits in one year? People who deny themselves every pleasure but money-making find that when they are ready to stop work and enjoy their fortune they have made a mistake. We have known poultry beginners, fanciful dreamers, to start with a plant costing as high as five thousand dollars, with fifteen hundred head of birds, and the stories of their failures were blazoned all over their districts. If the average person of either sex, of ordinary intelligence and ability, will start in a reasonable manner with ducks according to his or her means, and follow our teachings as given in this book, and by our letters, failure is next to impossible. An occasional mistake is not going to put one out of the business. An error can be corrected, and the lesson having been learned will not be repeated. There is positively no " out " about the duck business which will bring ruin. One man may not make as much money as another, that is to be expected. It is a question of starting with the right stock, following the right teachings and acquiring skill, experience and capacity, according to the individual.

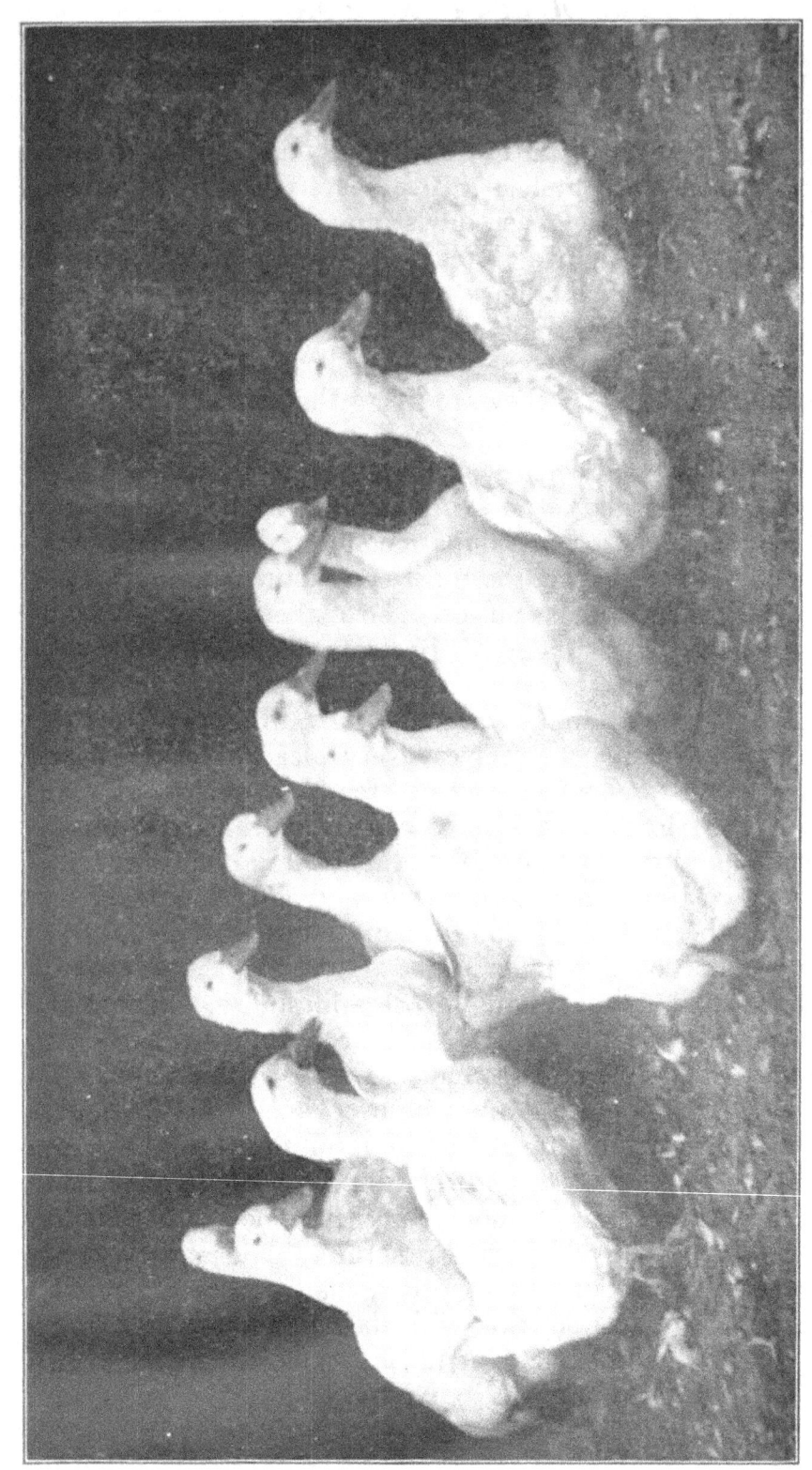

THE PARALLELOGRAM SHAPE OF BODY, DEPTH OF KEEL AND PLUMPNESS OF OUR DUCKLINGS ARE SHOWN HERE

QUESTION BOX

Q.—In my State (Texas) there do not seem to be any ducks like yours, and people with whom I talk do not think there would be any sale for them. *A.*—Texas folks like chickens, or beefsteaks. You don't live on cereals altogether, down your way, any more than we do. You have a clear field and, take our word for it, Texans will buy good ducklings, pay your price, and come back to you for more.

Q.—I have a solid gold watch for which I paid fifty dollars six months ago, and it is as good as new. How many ducks and drakes will you give me for it? I would like to be your agent out here, please quote agents' prices. *A.*—We sell for cash only; we have no agents and our prices are the same to everybody, to customers either in our own county or thousands of miles away.

Q.—If these ducklings are such exceptional eating, as you state, why can't I get up an attractive little booklet and circular and mail them around and get a list of steady buyers that way? *A.*—You can, and you ought to; that is just the way the finest eggs, poultry, butter, etc., are sold, at the highest prices.

Q.—I dislike the idea of killing the ducklings. Can I sell them alive? *A.*—Yes. There are thousands of poultrymen in the business of picking up birds alive from farmers, and marketmen everywhere who will take them alive.

Q.—I do general farming. Do you think I can grow the foods your ducks eat and do better feeding them to ducks than selling them separately? *A.*—Yes, you can get much more for them in the form of duckmeat. If breeders make big profits with ducks by buying everything, as we do, those farmers who have to buy only a part of the ration, raising the most of it, will make more money.

Q.—Instead of buying so much food stuff, why don't you raise it on your farm? *A.*—We wish to have our time free for the ducks and do not care to be busy at general farming, as we are satisfied with the profit on ducks from feeding bought food. We grow vegetables and green stuff for our needs, as that is no trouble.

HOW WE MAKE DUCKS PAY

Q.—**What is the proper time for killing ducklings?** *A.*—From ten weeks to three months of age. Keep them longer and they eat off the profits every day. The average farmer or housewife raising ducks for amusement by guesswork does not realize that. The age of ten to twelve weeks represents the maximum of plumpness and tenderness with the minimum of expense. After that age, the plumpness and tenderness decrease and the cost of keep is being added to all the time.

Q.—**I read what you say, that worms are fine for ducks and save on the grain bill. Why not raise worms? Is any treatise published on propagating worms? Guess you will think I am joking.** *A.*—Don't know of any worm guide. Would advise that you try the method seen in Belgium, where there is considerable swampy land. It is not uncommon there to find a yokel walking in the swampy land at the head of a flock of ducks, his wooden shoes at every stride squeezing out of the muck worms which the ducks gobble greedily—a sort of automatic self-feeding, non-paying scheme. What Yankee can beat this?

Q.—**Is there anything gained by breeding a small duck? A relative of mine living near here who did some duck breeding a few years ago used to get a good price for small ducklings and said his dealer sold them quicker than bigger ducklings.** *A.*—About ten years ago, some hotel and restaurant men in New York went to their duck and poultry markets and said: "We don't want you to give us too large a duckling. We get a good price now on our bills of fare and we wish from you (so we can make all the money possible) a medium-sized bird. When half of a medium-sized bird gets onto the table, it certainly will be large enough for one or two people. A whole big duckling is large enough for a family dinner. Keep the size down so they will cost us less and make us more money." The duck marketmen evidently did not argue the matter much. They passed the word along to the breeders in the territory south and west of New York, where most of the ducks then were being produced, and

QUESTION BOX

these breeders, to please their trade, actually began to breed for smaller size. Before long they were shipping to market ducks weighing not over four pounds. The result was what any sensible man could have predicted. The consumers, the men and women who ate the ducklings, complained and would not order them so often at hotels and restaurants. The hotel and restaurant managers soon changed their requests to " Give us the big ducklings again." The breeders went back to the early methods. They found building up size not so easy as pulling down. Qualities cherished and perpetuated by years of study and skill had gone. Today the flocks of those breeders are not yet back to a big-sized duckling, but feel the effects of the period of stunting. It is safe in poultry breeding to work for the biggest and juiciest. Anybody in any occupation who trades in an inferior article, hoping to make a bigger profit by selling it at the price of something better, is not playing fair to himself or to anybody.

Q.—**I live on the sea-coast. Is there anything in the air that would be prejudicial to the duck industry?** *A.*—Snow does not last long on sea-coast land, and this is a point in favor of the coast, because the ducklings will get out on the ground earlier in the spring, enjoy more exercise and do better. There is nothing in salt air or sea-breezes unfavorable to ducks.

Q.—**I see you have plenty of windows in your houses. How many are advisable?** *A.*—Put in windows freely. Light and sun are good for ducklings. When warm weather comes the windows are raised or taken out altogether so as to give plenty of free air.

Q.—**My ducks like to play in the muddy and swampy land. Will it hurt them?** *A.*—No, it will do them good. You can't keep them out of the muck. They will run for it.

Q.—**How will I get my ducks to lay?** *A.*—Feed them as we tell in the chapter on food. It is all a matter of food. They cannot lay unless they are nourished. If you starve them they will not do much for you.

HOW WE MAKE DUCKS PAY

Q.—**My ducklings do not seem very bright. They walked into a hole in the field and fell in and could not get out. Some were lamed and injured before I got them out.** *A.*—Ducklings are more or less stupid and must not be given a chance to fall into holes, or to run against sharp obstructions.

Q.—**I have read in poultry books that ground over which fowls run becomes tainted and unhealthful in time if something is not done to purify it. Is this true of a duck farm?** *A.*—Yes, and that is the reason crops of green stuff, like rye, are grown, to sweeten the soil. They do it, too, unfailingly.

Q.—**In dressing my first duck for the table, I tried to find the crop, to see if there was any food in it, but could not find it.** *A.*—A hen has a crop, but a duck has not. The food passage in a duck runs from the mouth to the gizzard.

Q.—**I have an incubator for hens' eggs. Can I use it for ducks' eggs?** *A.*—Yes. Of course, the ducks' eggs being larger, you cannot put in so many.

Q.—**Why is an incubator house built like a cellar?** *A.*—Because it will not be freezing cold there in the winter, and in summer it will be twelve to eighteen degrees cooler than outdoors (making it a good place to keep eggs then). To built an incubator cellar, dig only three feet deep, and use the dirt to bank up the walls outside. Build the walls of stone and cement. Put a roof over it, and a door at the end. Ventilate it well. Don't forget the ventilators, because egg-shells are porous. Dead air, with a large proportion of carbonic acid gas, and little oxygen, will influence badly the ducklings growing in the eggs.

Q.—**Why do you keep ducklings of different ages separate? Why not turn all in together, the young and the old?** *A.*—Because the old and strong birds would trample on and kill the smallest and weakest ones.

Q.—**At what age can ducklings be put out doors safely?** *A.*—When they are six weeks old, rain and

QUESTION BOX

cold will not hurt them, and they can be left out in their yards all night, unless it is bitter cold and stormy, and they will thrive better for it.

Q.—**What are your prices for breeding stock?** *A.*—See the sheet of prices accompanying this book. If the price-list is missing, or has been lost, write us for one, addressing your letter as follows to the Boston office: American Pekin Duck Company, 145 Pearl street, Boston, Mass.

Q.—**At what age does a duck begin to lay?** *A.*—When she is from four and one-half to five months old.

Q.— **Don't you lose any ducklings after you have hatched them?** *A.*—There is a loss of only about two out of every hundred.

Q.—**Can you ship live breeding ducks and drakes safely to me? I live in California.** *A.*—Yes; we ship the live birds by express any distance safely. We crate them lightly and strongly. For a long trip, a small bag of grain is sent along for food on the journey. The expressmen feed the birds and also water them. (There is a water tin attached to the crate.) On a long railroad run, the express agent, we find, gives the birds plenty of attention. It helps him pass the time away in the car, and he enjoys watching the birds.

Q.— **Do not the express companies make an extra charge for carrying your ducks a long distance?** *A.*—No. Live ducks and drakes go long distances at the single, or lowest, merchandise rate. For short distances, the one and one-half rate is charged (not the double rate, as for some animals).

Q.—**What is the color of eggs laid by your strain?** *A.*—White.

Q.—**If I buy some ducks and drakes of you, can I get along without an incubator?** *A.*—If you try to work without an incubator, you must have hens to set on and hatch the ducks' eggs. The ducks are not broody and will not set on and hatch their eggs. Better have an incubator.

FATTENING SHED AND PENS

As shown, the shed is open on both sides, giving plenty of fresh air at all times. It is simply a shelter from the sun, for the ducklings which are being fattened for market do not mind the rain at this age. The roof is not shingled. The cracks between the boards are covered with two-inch stripping. (No roofing paper here.)

The yards are separated by wire netting, but under the shed boards are used. The feed boards and water pails are shown in the yards. The pails are near the food boards so that the ducklings can waddle quickly from food to drink, and back again.

QUESTION BOX

Q.—**In case I do use a hen, how many duck eggs shall I put under her?** *A.*—Nine under a small hen, and eleven or twelve under a large hen.

Q.—**Will a hen brood the young ducklings, or would you provide a brooder?** *A.*—Get a brooder. A hen is awkward in brooding ducklings, as a rule, injuring some by crushing.

Q.—**How would you advise working up a duck market in a place where ducklings are comparatively unknown?** *A.*—Make the prospective customer a gift of a duckling and let him or her serve it for dinner. He will be won over by the experiment and, we predict, will report to you that the dish is ahead of chicken or turkey. Sales will follow as a matter of course. This is a good way to get acquainted with a hotel or restaurant keeper.

Q.—**Is your system of feeding followed by many?** *A.*—There are some breeders of ducks who care more for low cost than they do for flavor of meat. They feed fish caught from the ocean or lakes in nets. Ducklings fattened on fish taste fishy and their flesh is not fine-grained.

Q.—**Would you advise a woman to go into duck raising?** *A.*—It depends on the woman, as it does on the man, also. Some men get enthusiastic, and like to figure the money they are going to make, but after a while cool off and become lazy and indifferent. A woman who likes hens and chickens will like ducks. The work is not much different. We know of good work done by women raising ducks. Of course women should hire help if they run a large duck farm. Some women are better fitted for poultry raising than men.

Q.—**What would you advise with regard to the selection of a farm?** *A.*—No matter how poor and how cheap the land, the manure from the ducks will fertilize it. Land which has a gentle slope, or which is gravelly, will be drained better than low, level land. It ought not to be possible for pools of water to form and get stagnant.

Q.—**Please explain more fully what a beginner should**

guard against when buying breeding stock. *A.*—It is not uncommon for a beginner with ducks to buy breeding stock of unknown sources. Ducks and drakes may be offered at a low price in market stores or by unknown and irresponsible breeders. **The novice should be very much on his guard against such stock.** Unless you can start with reliable stock bred skillfully under proper conditions, it is better not to start at all. A few years ago a gentleman, known in New England now as the manager of a stock farm, experimented with ducks and learned by sorrowful experience what we are telling you here. He gives the story in his own words as follows:

" We started with four large-sized incubators and about one hundred and eighty Pekin ducks. Foods were figured down to percentages, and we had to study arithmetic all over to learn what the ducks should have for supper, and tackle algebra to mix a proper breakfast. This was not the fault of our teacher, but was about the type of the general instructions for the feeding of poultry, and it is not yet all done away with in some quarters. We filled the first incubator with **four hundred and eleven eggs. We managed to hatch seventy-eight.** Our second hatch was **one hundred and thirty-four out of four hundred and seventy-one eggs.** Third hatch was **one hundred and sixty-five out of five hundred and twenty eggs. About one hundred died after pipping.** The hatches grew steadily worse, and the ducklings correspondingly weaker. A great many saw with one eye and walked with one leg. We asked the authorities, and found that our food was too hot, and that it was too cold; we fed too much meal, and too little meal; we did not ventilate enough, and they had too much air; we ought not to give them too much grit, and we must be sure to give them plenty; we must keep the house warm, and again it should not be very warm for ducks. They felt sure the temperature in the machine was too high or too low, and the ventilation was both ways. **We did not find one who put the trouble where it belonged — in the eggs — and, back of that, in the breeding stock.** The next year we got as low as **seventy ducklings** hatched from a full machine of eggs (**four hundred and fifty to five hundred**). We found out that we had been wasting our time with poor stock as breeders. We got together another flock and this time did better. **To produce strong, healthy ducklings one must have vigorous, hardy breeding ducks that have been selected and fed to that end, and**

QUESTION BOX

not, as is so often the case, merely chosen from a hap-hazard lot of forced market birds and thrown out as being good enough for breeders."

The above should be read carefully and remembered. Get your breeding ducks of us and you will have hardy, vigorous birds that can lay fertile eggs. Do not start with unknown stock or stock of doubtful value offered at a low price and go through the worries and vexations described in the foregoing paragraph.

SHELTER ROOF FOR DUCKS ON THE RANGE

www.ingramcontent.com/pod-product-compliance
Lightning Source LLC
Chambersburg PA
CBHW082336220526
45470CB00008B/2530